Physics of Dielectric Solids, 1980

Physics of Dielectric Solids, 1980

Invited papers presented at the Conference on Physics of Dielectric Solids held at the University of Kent at Canterbury, 8–11 September 1980

Edited by C H L Goodman

Conference Series Number 58

The Institute of Physics
Bristol and London

This work relates to Department of the Navy Research Grant N00014-80-G-0070 issued by the Office of Naval Research. The United States Government has a royalty-free license throughout the world in all copyrightable material contained herein.

CODEN IPHSAC 1–151

British Library Cataloguing in Publication Data
Conference on Physics of Dielectric Solids (*1980: University of Kent*)
Physics of dielectric solids, 1980. – (Conference Series/Institute of Physics;
ISSN 0305-2346; no. 58)
1. Dielectrics – Congresses
I. Title   II. Goodman, C H L
III. Institute of Physics
537.2'4   QC584

ISBN 0-85498-149-7
ISSN 0305-2346

The Conference on Physics of Dielectric Solids was organised by the Solid State Sub-committee of The Institute of Physics, in collaboration with The Institution of Electrical Engineers and The Dielectrics Society under the sponsorship of the European Research Office (USARSG)/European Office of Aerospace Research and Development (EOARD), the Department of the Navy, Office of Naval Research, Arlington, Virginia, the Central Electricity Generating Board and ICI Limited.

Programme Committee
C H L Goodman (Chairman), A R Blythe, G C Chantry, R M Hill, A M North, W A Phillips, G Williams

Organising Committee
A C Lynch (Chairman), A K Jonscher (Conference Secretary), C H L Goodman (Programme Committee Chairman), A D Buckingham, J H Calderwood, J Champion, G W Chantry, D K Davies, J Gibbings, T J Lewis, G J Morgan, G G Roberts, E W Williams, G Williams

Conference Chairman
H Frölich FRS

Honorary Editor
C H L Goodman

Published by The Institute of Physics, Techno House, Redcliffe Way, Bristol BS1 6NX, and 47 Belgrave Square, London SW1X 8QX, England.

Set in 10/12pt Press Roman by DJS Spools Ltd, Horsham, Sussex and printed in Great Britain by J W Arrowsmith Ltd, Bristol.

Preface

'The Solid State' is often talked of as one of the best understood and most highly perfected branches of Physics. While that may be arguable in the case of semiconductors or, perhaps, metals, it is certainly not true of dielectrics. These, of course, are the earliest solids to have attracted man's scientific curiosity – from the Stone Age amber has been treasured for its unusual properties, and electricity is named from the Greek word for it. However, although it is obvious that dielectric solids have been the subject of detailed study for at least a century, a major problem in arriving at any unified understanding of their physical behaviour has been the vast amount of data that has been collected – a mass with rather few informing principles. How, then, does one see the wood with so many trees in evidence?

It was, in fact, a discussion on very much these lines, in the Institute of Physics Solid State Physics Subcommittee, that led to the idea of a Conference on the Physics of Dielectric Solids. It was many years since a conference had been devoted to anything of the kind, and developments in other fields of Solid State Physics could perhaps help gain better understanding of dielectric behaviour. That led to the Institute holding a small meeting of specialists in various fields of dielectric research, who enthusiastically supported the suggestion. An organising committee was formed, Professor H Fröhlich, the doyen of the field, gave his blessing, as did the Institution of Electrical Engineers and the Dielectrics Society, and the eventual result was the fascinating conference held at the University of Kent at Canterbury, 8–11 September 1980.

From the start the aim was twofold: tutorial invited papers setting specific themes in what were felt to be key areas, and very adequate time for discussion of related contributed papers in poster sessions. Deliberately, there were no parallel sessions.

The invited papers only are included in the present volume, and, I hope, give something of the flavour of the Conference, and its intentional mixing of new and classic themes. Obviously, however, these set pieces can give little clue as to the warmth of discussion and the absorbed involvement of all who attended, or how the bringing together of so many different points of view may have generated some heat, but much light! Though not so styled, it was indeed an international conference in that more than half the delegates came from abroad; and it was strongly felt by all those present that a further meeting of the kind should be planned for, say, three years on.

Although it is invidious to name names, since so many contributed much time and enthusiasm in planning the Conference and its programme, a particular debt is due to Dr Arnold Lynch, who chaired the Organising Committee and Professor Andrew Jonscher, the Conference Secretary – and of course to the traditionally anonymous referees who sorted out many intriguing points in the present volume.

C H L Goodman

Contents

Inst. Phys. Conf. Ser. No. 58
Invited paper presented at Physics of Dielectric Solids, 8–11 September 1980, Canterbury

Molecular correlation function approaches to dielectric relaxation

R H Cole
Department of Chemistry, Brown University, Providence, RI 02912 USA

Abstract. After a discussion of formal linear response theory of time-dependent polarisation and brief indication of the current state of equilibrium theory, the conditions for the time evolution operator to give Debye relaxation are considered, with examples of this (infrequent) behaviour in solids. A review of empirical and semi-empirical relaxation functions with discussion of their implications for molecular dynamics and examples is followed by an examination of more specific theoretical models and summary of results using Brownian dynamics, Monte Carlo simulations and stochastic analytical methods. The point is made that these diverse approaches lead to deviations from exponential behaviour which are remarkably similar, at short times especially, as a result of cooperative many-body interactions on the molecular scale. Finally, relations between dipole and other observable correlation functions are considered briefly, as are possibilities for relating nonlinear electric field effects to correlation functions.

1. Introduction

The problem of understanding why dielectric relaxation in condensed phases of matter so often deviates in characteristic ways from simple exponential (Debye) behaviour has been a real one for at least a century. The writer has been concerned with it for more years than he cares to remember, and for most of the time has had an intuitive feeling that in most cases the answer or answers must lie in many-body treatments of the workings of intermolecular forces, rather than in treatments based on mean-field models and heterogeneity of the system. Only recently, however, has this belief been solidly reinforced by results based on reasonably realistic and detailed molecular models.

Most of the recent theoretical developments work directly or indirectly toward better solutions by use of the formalism of linear response theory and evaluation of correlation functions appearing as a result of the formalism. The invitation to present an account of 'the molecular correlation function approach to dielectric relaxation' has given me a welcome opportunity to review the current state of progress in understanding dielectric relaxation in these terms. The plan of the paper is outlined very briefly in the abstract. It seems best not to attempt a more adequate summary in this introduction, but rather to proceed directly to the more detailed, but still far from complete, account.

2. Formal correlation function theory

The macroscopic electric polarisation, or moment density, $\mathbf{P}_\mu(t)$ of electric dipoles μ_i

0305-2346/81/0058-0001 $01.00

in a volume V can for our purposes be expressed as

$$\mathbf{P}_\mu(t) = (1/V)\langle \Sigma_i \mu_i \rho(t) \rangle \tag{1}$$

where the time-dependent distribution function $\rho(t)$ satisfies the Liouville equation $\partial\rho/\partial t + L\rho = 0$, and an ensemble average is taken. The classical Liouville operator L as a function of coordinates q_i and conjugate momenta p_i is related to them and the Hamiltonian H of the system by

$$L = \Sigma_i \left(\dot{p}_i \frac{\partial}{\partial p_i} + \dot{q}_i \frac{\partial}{\partial q_i} \right) = -\Sigma \left(\frac{\partial H}{\partial q_i} \frac{\partial}{\partial p_i} - \frac{\partial H}{\partial p_i} \frac{\partial}{\partial q_i} \right) \tag{2}$$

In linear response theory of dielectrics (Kubo 1957, Glarum 1960), the linear effect of an applied field $E(t)$ in the z direction in producing a perturbation $\rho_1(t)$ from the equilibrium distribution ρ_0 in no field, with L_0 and H_0 the corresponding Liouville operator and Hamiltonian, is given by

$$\frac{\partial \rho_1}{\partial t} + L_0 \rho_1 = E(t)\, \Sigma_i \frac{\partial \mu_{zi}}{\partial z_i} \frac{\partial \rho_0}{\partial p_{zi}} = -E(t)\, (\Sigma_i \dot{\mu}_{zi})\, \rho_0 / k_\mathrm{B} T \tag{3}$$

where $\rho_0 = A \exp(-H_0/k_\mathrm{B}T)$. Formal solution for $\rho_1(t)$ gives the time-dependent part of $\mathbf{P}_z(t)$ from equation (1) as

$$\mathbf{P}_{\mu z}(t) = -(k_\mathrm{B}TV)^{-1} \int_{-\infty}^{t} \mathrm{d}t_1 E(t_1) \langle (\Sigma_i \mu_{zi})\, \mathrm{e}^{-(t-t')L_0}\, (\Sigma_i \dot{\mu}_{zi}) \rho_0 \rangle . \tag{4}$$

The central problem posed by this approach is evaluation of the dipole correlation function $\Phi_{\mu z}(t)$:

$$\Phi_{\mu z}(t) = \langle (\Sigma_i \mu_{zi})\, \mathrm{e}^{tL_0}\, (\Sigma_i \mu_{zi}) \rho_0 \rangle = \langle (\Sigma_i \mu_{zi})\, (\Sigma_i \mu_{zi}(t)) \rho_0 \rangle \tag{5}$$

where, in the second form, use has been made of the fact that $\mu_{zi}(t) = \exp(tL_0)\, \mu_{zi}$ is the formal solution of the equation $\mathrm{d}\mu_{zi}/\mathrm{d}t = L_0 \mu_{zi}$ for the 'natural motion' of dipole μ_{zi} in the absence of the field.

Because equation (4) has the form of a convolution, the solution for response to an alternating field $E(t) = E \exp(\mathrm{i}\omega t)$ is obtained immediately from the Laplace transform $\mathscr{L}_{\mathrm{i}\omega} \mathbf{P}_{z\mu}(t)$, giving the dipole contribution $\epsilon^* - \epsilon_\infty$ to the complex relative permittivity ϵ^* as

$$\epsilon^* - \epsilon_\infty \sim (k_\mathrm{B}TV)^{-1} \mathscr{L}_{\mathrm{i}\omega} [-\dot{\Phi}_{\mu z}(t)] = (k_\mathrm{B}TV)^{-1} [\Phi_{\mu z}(0) - \mathrm{i}\omega \mathscr{L}_{\mathrm{i}\omega} \Phi_{\mu z}(t)] . \tag{6}$$

The equilibrium or static permittivity $\epsilon_\mathrm{s} - \epsilon_\infty$ is thus determined by the equilibrium correlation function $\Phi_{\mu z}(0)$ and the frequency dependence by $\mathscr{L}_{\mathrm{i}\omega} \Phi_{\mu z}(t)$, which can be written formally as

$$\mathscr{L}_{\mathrm{i}\omega} \Phi_{\mu z}(t) = \langle (\Sigma_i \mu_{iz}) \frac{1}{\mathrm{i}\omega + \mathscr{L}_0} (\Sigma \mu_{iz}) \rho_0 \rangle . \tag{7}$$

This form provides an alternative to equation (5) as a statement of the dynamical problem which can be a useful starting point, as used by Zwanzig (1963) for example.

In the time domain $\Phi_{\mu z}(t)$ is obtained directly from the decay of polarisation following removal of a constant field at $t = 0$, and $\mathrm{d}\Phi_{\mu z}/\mathrm{d}t$ from the transient discharge current, with simple counterparts for response to a constant field applied at $t=0$. In such expressions the effects of polarisation, which is 'instantaneous' on the time scales of permanent dipole displacements and represented by ϵ_∞, appear as a step function at $t=0$.

Before considering general aspects of the time-dependent polarisation problem as expressed by equations (5) or (7), a brief digression to comment on the equilibrium problem is appropriate. The equilibrium dipole correlation function $\Phi_{\mu z}(0)$ appears almost casually in equation (6), but the problem of evaluation to obtain $\epsilon_s - \epsilon_\infty$ for even moderately polar substances is at present in a less well defined and satisfactory state than it seemed to be ten years ago. Then it appeared that Onsager's equation was a reasonably good relation between molecular dipole moments and static permittivities, with substantial deviations only for strong dipole interactions or specific short range effects such as association and hydrogen bonding. Since then, several statistical many-body dipole interaction theories have been developed, all of which predict static permittivities much larger than the Onsager value for even relatively small values of dipole interaction strength as expressed by the variable $y=\mu^2/k_BTr^3$ where r is an intermolecular spacing (see Wertheim (1971) and Patey (1977), for example).

The extent of the deviations can be suggested by the fact that the value of y appropriate to liquid water gives permittivities larger than the experimental one, ordinarily accounted for by a Kirkwood g factor of order 2.5 as a result of intermolecular hydrogen bonding. These theories of point dipole interactions have been developed primarily for polar liquids, but calculations by the writer (Cole 1977) for the spherical model of dipole–lattice interactions developed by Toupin and Lax (1957) give very similar results. The moral seems to be that the point dipole interaction models used are a poor representation of real molecular charge interactions and orientation-dependent intermolecular forces in such systems, and calculations by Patey showing that addition of a molecular quadrupole force drastically decreases the predicted permittivities to values more in line with observed behaviour are consistent with this surmise.

If one imposes an uncompromising stricture that calculations of the time-dependent $\Phi_{\mu z}(t)$ can be no better than the results for the simpler $\Phi_{\mu z}(0)$ to which they reduce, all the discussions of $\Phi(t)$ to follow must be regarded with considerable suspicion. We shall take the more optimistic view that the uncertainties of these local, or even internal, field problems are reflected primarily in magnitude rather than form of time or frequency dependences.

Returning to the formal expressions for $\Phi_{\mu z}(t)$, one recognises first that strictly L_0 defined classically by equation (2) is a many-body operator in terms of all coordinates and momenta, with all the interactions present in the Hamiltonian H_0 of the entire system. Many of these will be inactive or ineffective for the time evolution dynamics of a given dipole, either directly or in ensemble average. This can, it is hoped, make calculations for simplified models reasonably realistic as well as tractable, but as in any macroscopic result the observable dielectric relaxation behaviour can reflect and serve as a probe of only limited aspects of the molecular dynamics. As compared to other equally limited but different probes, dielectric measurements do have the advantage of a wider range of times and frequencies which are accessible with varying degrees of difficulty.

Most of the discussion to follow will be concerned with molecular dipole correlations, but it should be recognised that the response theory and correlation function formalism can be developed and is useful for charge transport more generally. If one has both molecular dipoles μ_{zi} and displacements z_i of charges e_i, only the macroscopic total current J_t expressed by

$$J_t = V^{-1}\langle\Sigma_i(\dot{\mu}_i + e_i\dot{z}_i)\,\rho(t)\rangle$$

is observable by purely electromagnetic measurements. The response theory formation then gives as counterpart of the dipole correlation function $\Phi_{\mu z}(t)$ the current or velocity correlation function $\psi(t)$

$$\psi(t) = (k_B TV)^{-1}\langle\Sigma_i(\dot{\mu}_{zi} + e_i\dot{z}_i)\,e^{tL_0}\Sigma_i(\dot{\mu}_{zi} + e_i\dot{z}_i)\,\rho_0\rangle. \tag{8}$$

Effects of distinct mobile charge carriers can thus be expressed in terms of self and joint correlations of their velocities and there may also be cross correlations with dipole 'currents' $\dot{\mu}_{zi}$, as in kinetic depolarisation and dielectric friction effects.

Our discussion will further be primarily for systems and time scales such that classical dynamics is adequate and appropriate. The usual conditions for this to be so are that the frequencies ω of interest satisfy $\omega \ll k_B T/\hbar$, or that negligible motion occurs in times of order $t = h/k_B T$. Even when these conditions are not satisfied, however, evaluation of the classical limiting behaviour may still be adequate if a quantum statistical factor is included. The domains for the conditions just stated to be satisfied at room temperature roughly restrict frequencies to less than 100 GHz and times to greater than 1 ps.

3. Debye relaxation; discrete distributions

The simplest form of dielectric response, and a prototype for many discussions, is the behaviour when individual dipole moments $\mu_{zi} = \cos\theta_i$, for angle θ_i between the dipole and the field, are eigenfunctions of the Liouville operator L_0, $L_0\mu_{zi} = -k\mu_{zi}$. In the time domain one then has $\Phi(t) = \Phi(0)\exp(-kt)$ and in the frequency domain

$$\epsilon^* - \epsilon_\infty = (\epsilon_s - \epsilon_\infty)\,\frac{1}{1 + i\omega\tau_D} \tag{9}$$

where $\tau_D = 1/k$ is the relaxation time. We shall follow common usage in referring to this form of time or frequency behaviour as Debye relaxation without implying any particular molecular mechanism or interpretation of τ_D. Debye (1929) obtained this result for a rotational diffusion model of reorientations of spherical molecules in a viscous medium with $\tau_D = 3\eta V/kT$, but the same form of time dependence results from *any* model in which the probability of transitions between two orientations is a time-independent function $k(\varphi)$ of the angle φ between orientations only, with τ_D given by

$$1/\tau_D = \int d\varphi\, k(\varphi)[1 - \cos\varphi].$$

This result, first pointed out by Van Vleck and Weisskopf (1945) and readily obtained with the correlation function formalism (Cole 1965), is a simple example of the truism to which we return later: that an observed form of kinetics is no proof of a particular model predicting that form.

Although relaxation of Debye form is often a good approximation to the behaviour of liquids or solutions containing simple polar molecules and at not too low temperatures, examples for solids are relatively few and as often as not rather special. A partial list, gathered from tabulations in Böttcher and Bordewijk (1978) and various other sources easily recalled by the writer, includes ice near its melting point, some 'soft' (non-ceramic) ferroelectrics, solid alcohols, and high-temperature phases of some aliphatic alcohols, hydrogen halides and substituted benzenes.

The form of Debye behaviour in time or frequency is shown in figure 1, as are commonly observed types of deviation discussed in later sections. We recall here the

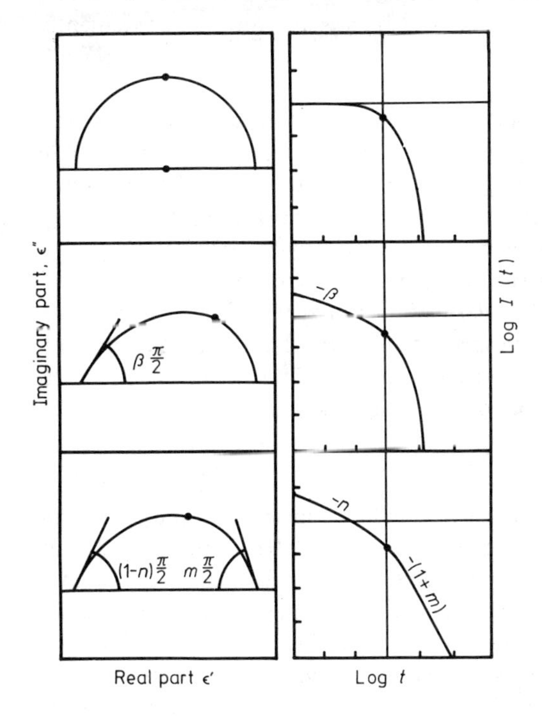

Figure 1. Left: Complex plane loci of permittivity ϵ^* for Debye (top), Cole–Davidson (centre), and Havriliak–Negami (bottom) empirical relaxation functions. Right: Logarithmic plots of transient currents $I(t)$ versus time t for the same functions.

familiar fact that the single Debye loss peak in ϵ'', which is symmetrical on a logarithmic scale about $\omega\tau_D = 1$, is very broad by absorption spectroscopy standards, as the half width at half height is 0.57 decades. Two Debye peaks with ratio of times τ_D less than four are thus unresolved by conventional linewidth criteria and distinguished with difficulty. Conversely, readily apparent deviations result only from broad distributions of relaxation times on a logarithmic scale, if this concept is invoked, or from significantly different forms of relaxation function which may be recognisable over several decades of time or frequency.

Sums or distributions of Debye relaxation functions when one alone is inadequate have been proposed, from almost the beginning of experimental information, by writing expressions of the form

$$\Phi_{\mu z}(t) = \Sigma_i A_i \exp(-t/\tau_i), \qquad \epsilon^* - \epsilon_\infty = \Sigma_i A_i \frac{1}{1 + i\omega\tau_i},$$

while for broad relaxations over several decades logarithmic distribution functions $g(\tau/\tau_0)$ have been introduced by such defining equations as

$$\epsilon'' - \epsilon_\infty = (\epsilon_s - \epsilon_\infty) \int d(\ln \tau/\tau_0)\, g(\tau/\tau_0) \frac{1}{1 + i\omega\tau}. \tag{10}$$

Jonscher (1975, 1975b) has concluded that equation (10) 'has little meaning' and that 'the physical significance of $g(\tau/\tau_0)$ has never been made very clear'. A more cautious statement is that there are some systems in which the existence of a few discrete functions has been given a satisfactory theoretical interpretation, that solutions of sets of coupled rate equations for particular models can produce the equivalent of a broad distribution, but that in many cases such distributions are neither a necessary nor even conveniently derivable form of prediction from theoretical models for comparison with experiment. Examples of each proposition follow.

There are several models, more appropriate in the main for liquids than for solids, in which two or more observed discrete relaxation functions can be interpreted in terms of diffusion-like motion of nonspherical molecules and internal as well as overall reorientations of dipole groups within flexible molecules. Models of a set of low-energy orientational sites for molecules in solids which are separated by potential barriers with different activation energies have often been proposed. A classic example of this kind of model with a real, rather than *ad hoc*, molecular basis for particular systems is provided by the work of Hoffman (1955) and Lauritzen (1958). Denoting the time-dependent probability of occupation of a given site i with transition probabilities k_{ij} to adjacent sites j by P_i, the relaxation is described by a set of coupled rate equations

$$\frac{\mathrm{d}P_i}{\mathrm{d}t} = \Sigma_j (k_{ji}P_j - k_{ij}P_i). \tag{11}$$

These admit solution for normal modes k as linear combinations of the P_i and relaxation functions $\exp(-t/\tau_k)$ determined by vanishing of the secular determinant for the P_i of equation (11). Lauritzen developed molecular models of the k_{ij} for long chain ketones and bromides reorienting among six-fold sites in tunnel-like cages of urea clathrate compounds, both from volume overlap arguments and use of Lennard–Jones 6–12 atom–atom potentials, and with use of these k_{ij} was able to predict multiple peaks observed by Meakins and Mulley (1954) with remarkable success. This is a case where barrier models have a sound physical basis from independent structural evidence, but it is important to recognise that such site models do not easily account for amplitudes of relaxation modes with greatly different relaxation times or for very many such modes. More likely situations which can be described by sets of coupled rate equations with quite different reasons for different transition probabilities are described in §6.

Until recently there have been considerable difficulties in merely fitting an empirical distribution of Debye relaxation function to available experimental data, but Colonomos and Gordon (1979) have presented a procedure worth mentioning here because it appears to provide a potentially valuable tool. They were able to show that by fitting experimental data for ϵ' and ϵ'' at N frequencies numerically with Tchebyshev polynomials, both upper and lower bounds can be placed on cumulant distribution functions (i.e., the integral or sum to a given τ of $g(\tau/\tau_0)$) which can fit the data. This is clearly a considerable advance over previous methods because one has the possibility of distinguishing a small number of discrete relaxation processes, when they exist, or of defining a profile or range in which continuous distributions from empirical or otherwise derived theoretical relaxation functions must fit if they can represent the data. Also, the upper and lower bounds, obtained by linear programming methods, can provide possibly quite sobering indications of how much can reliably be inferred from a given set of data and should lead to proper humility in this respect.

4. Empirical relaxation functions

Many empirical relations have been proposed to describe dielectric and other relaxation processes. We consider here only some of these which are frequently used and of interest for comparison with results of molecular models described in §§5 and 6. More complete descriptions of most of them have been given in the chapter on empirical descriptions of dielectric relaxation in Böttcher and Bordewijk (1978) with 358 references.

The listing here is roughly in order of increasing deviations from Debye behaviour with rather similar descriptions in order of their first appearance, and some examples of systems whose behaviour has been described by each are also given. As for the discussion of Debye relaxation, these have been gathered from a variety of sources, including Böttcher and Bordewijk, annual Digests of the Literature on Dielectrics, and personal recollection. No attempt has been made at completeness, or at assessments of accuracy of the data, quality of fit, and whether a different representation would be better. The reader should be cautioned that all too often results reported in the literature have been based on data for a very limited number of frequencies or times, and that results and conclusions of different investigators are sometimes not in good agreement.

4.1. The Cole–Davidson 'skewed-arc' function

In the frequency domain, this function has the form

$$\epsilon^* - \epsilon_\infty = (\epsilon_s - \epsilon_\infty)\,\frac{1}{(1 + i\omega\tau_0)^\beta}, \qquad (0 < \beta < 1). \tag{12}$$

It was originally proposed by Davidson and Cole (1950) to describe deviations from Debye behaviour on the high-frequency side of the loss region in supercooled glycols and has since been used more frequently for a variety of liquids and polymer solutions than for solids. A few examples of the latter are the rigid rotator phase of pentachlorotoluene, 1,3 dithiane below −50°C, 2,4,6 tri-*t* butyl phenol, butyl stearate, and patially disordered phases of solid hydrogen bromide and iodide just above their low-temperature solid-phase transitions (Groenewegen and Cole 1967).

The time and frequency behaviour of the skewed-arc function are sketched in figure 1. The frequency dependence is 'Debye-like' for $\omega\tau_0 \ll 1$ and a power law $\epsilon^* \sim (i\omega)^{-\beta}$ for $\omega\tau_0 \gg 1$ as so frequently observed in solids. The proposition that this sort of behaviour is more liquid-like than solid-like (Jonscher 1975a) is at least debatable, but the thesis that it is most commonly observed at frequencies above 10^7 Hz is certainly not borne out by the wealth of data for supercooled liquids and solutions with fewer maxima at frequencies down to 1 Hz and decreasing values of β at lower temperatures.

4.2. The Williams–Watts function

This expression for the macroscopic dipole correlation function or decay of polarisation after removal of an applied field at $t = 0$ is the real time function

$$\Phi(t) = \Phi(0)\exp\left[-(t/\tau_0)^{\beta_W}\right] \tag{13}$$

where, like the Cole–Davidson β, the parameter β_W has values in the range $0 < \beta_W < 1$.

The function was proposed by Williams and Watts (1970) as a representation of relaxation processes for supercooled solutions of a variety of solutes in o-terphenyl and decalin. The complex plane plots of ϵ^* (obtained by series expansion of equation (13) and Laplace transformation) resemble those for the skewed-arc function, but for the same asymptotic high-frequency behaviour, the Williams–Watts expression gives a smaller loss maximum with deviations from Debye behaviour apparent at lower frequencies. For β_w less than about 0.5, there is a considerable range of frequencies below $\omega\tau_0 = 1$ for which the loss varies approximately as a fractional power of frequency, $\epsilon'' \sim (\omega\tau_0)^{1-m}$, even though ϵ'' varies asymptotically as ω for $\omega\tau_0 \ll 1$.

In addition to its original uses, the Williams–Watts function has been employed to represent relaxation processes in a number of amorphous polymers with large deviations from Debye behaviour ($\beta_w < 0.5$), but it seems to be true that in some cases better fits can be obtained using the Havriliak–Negami equation discussed in §4.5.

4.3. The Cole–Cole 'circular-arc' function

This widely used relation had its origin as a representation of electrical impedance properties of biological membranes and tissues (Cole 1928). In a search for analogies in what were hoped to be simpler physical systems, the counterpart of the impedance function was tested as a representation of available data on liquids and solids by K S Cole and R H Cole (1941).

The simplest expression of the circular-arc behaviour is by the equation for the complex permittivity

$$\epsilon^* - \epsilon_\infty = (\epsilon_s - \epsilon_\infty)\frac{1}{1 + (i\omega\tau_0)^{1-\alpha}}, \qquad (0 < \alpha < 1) \tag{14}$$

and the complex plane plot is the familiar circular arc with depressed centre rather than the semicircle for Debye behaviour. The ϵ' and ϵ'' curves as a function of frequency are symmetric on a logarithmic scale of frequency, with the loss varying as $(\omega\tau_0)^{-(1-\alpha)}$ for $\omega\tau_0 \gg 1$ and as $(\omega\tau_0)^{+(1-\alpha)}$ for $\omega\tau_0 \ll 1$. The latter is of course in contrast to the approach to Debye behaviour for $\omega\tau_0 \ll 1$ of the asymmetric Cole–Davidson and Williams–Watts functions. The asymptotic time-dependences of $\Phi(t)$ (from series expansions of equation (14) and Laplace transformation) are of the form $(t/\tau_0)^{-\alpha}$ for $t/\tau_0 \ll 1$ and $(t/\tau_0)^{-(2-\alpha)}$ for $t/\tau_0 \gg 1$.

The circular-arc function has been used, sometimes on the basis of fragmentary data, to describe relaxation of many liquids and solids, the latter often of dubious and uncertain composition. A few examples of solids for which the data seem reasonably reliable are several crystalline substituted benzenes, the hydrogen halides HBr and DBr in their ordered low-temperature phases, furan, and formic acid. The function has also been used to describe relaxation of a variety of polymers in solution and as amorphous solids, but as already mentioned it appears that the Havriliak–Negami function is often a better description.

4.4. The Fuoss–Kirkwood function

The behaviour of this function, proposed by Fuoss and Kirkwood (1941) at the same time as the Cole–Cole relation for dielectric behaviour, quite closely resembles the latter, despite a somewhat different functional form. Their expression for ϵ'' as a function of ω has the attractively simple form

$$\epsilon''(\omega) = \frac{2\epsilon''_m}{(\omega\tau_0)^{-\alpha^1} + (\omega\tau_0)^{+\alpha^1}} \tag{15}$$

where ϵ''_m is the maximum loss at $\omega\tau_0 = 1$, and $0 < \alpha^1 < 1$. This is again logarithmically symmetrical about $\omega\tau_0 = 1$. For $\alpha^1 = 1 - \alpha$ the asymptotic frequency dependences are the same as for the circular-arc function, but the loss maximum is somewhat larger, especially for small values of α^1.

The Fuoss–Kirkwood function was originally proposed as an empirical generalisation of a theoretical result by Kirkwood and Fuoss (1941) for relaxation of chain polymers undergoing conformational rearrangements in a solvent treated as a hydrodynamic continuum. Interestingly in the context of the discussion of relaxation time distributions above, Kirkwood and Fuoss obtained their solution in terms of a theoretically derived $G(\ln\tau)$ of relaxation times of the form

$$G(\ln\tau) = \frac{\tau_0}{(\tau + \tau_0)^2}, \qquad \tau > \tau_0$$

where τ_0 is a relaxation time for polymer segment motions.

Although different, the Fuoss–Kirkwood and Cole–Cole predictions are similar enough that a choice between them in fitting a given set of data is difficult unless the precision is good over an extended frequency range. It is probably true that many data reported in the literature as fitted by one could be fitted as well by the other, but in a few cases studied by the writer, the circular-arc function could be adjusted for a better fit (see, e.g., Cole (1955)).

4.5. The Havriliak–Negami equation

Havriliak and Negami (1966, 1967) proposed their equation to fit data on amorphous polymers and polymer solutions which were intermediate in character between the Cole–Davidson or Williams–Watts behaviour on the one hand and the circular-arc or Fuoss–Kirkwood descriptions on the other. Their relation for the complex permittivity is, so to speak, a combination of the Davidson–Cole and Cole–Cole functions:

$$\epsilon^* - \epsilon_\infty = (\epsilon_s - \epsilon_\infty)\frac{1}{[1 + (i\omega\tau_0)^{1-\alpha}]^\beta}. \tag{16}$$

With the empirical parameters α and β in the range 0 to 1, an asymmetric complex phase locus and logarithmic frequency dependence of ϵ' and ϵ'' results with the loss ϵ'' varying as $(\omega\tau_0)^{-(1-\alpha)\beta}$ for $\omega\tau_0 \gg 1$ and as $(\omega\tau_0)^{1-a}$ for $\omega\tau_0 \ll 1$. The forms of the time and frequency dependences are shown in figure 1.

The principal use of this function has been for polymers and it seems clear that it can fit results for many amorphous polymers and polymer solutions very satisfactorily over considerable ranges of frequency. A remarkable example is from the recent results of Yoshihara and Work (1980) for atactic poly(4-chlorostyrene) at temperatures from 406 to 446 K and 0.2 Hz to 0.2 MHz. The deviations of their very precise data from a best fit with $\alpha = 0.136$, $\beta = 0.515$ in equation (15) did not exceed 0.2 per cent of $\epsilon_s - \epsilon_\infty$ and were systematic, possibly as a result of mechanical modulations of the cell. Interestingly, results of much the same kind, but with necessarily lower precision, were obtained by Mashimo (1976) for solutions of poly(4-chlorostyrene) in non-polar solvents exhibiting relaxation in the megahertz range.

4.6. The Jonscher and Hill functions

Jonscher (1975a) first proposed an empirical relation for the loss ϵ'' as a function of frequency of the form

$$\epsilon'' = \frac{A}{(\omega\tau_0)^{-m} + (\omega\tau_0)^{1-n}} \tag{17}$$

with the parameters m and $1-n$ in the range 0 to 1. This can be regarded as a generalisation of the Fuoss–Kirkwood equation (15) to which it reduces for $m = 1-n = \alpha^1$. For m different from $1-n$ it obviously yields an asymmetric loss peak as a function of log $\omega\tau_0$ with limiting power law dependences as $(\omega\tau_0)^{-(1-n)}$ for $\omega\tau_0 \gg 1$ and $(\omega\tau_0)^m$ for $\omega\tau_0 \ll 1$. Hill (1978) has reported analysis of loss data for some 50 substances, mostly solids, for which values of $1-n$ and m could be chosen to fit the data, and for all but eight found $1-n$ was equal to or greater than m, a behaviour resembling that given by the Havriliak–Negami equation, with the other empirical equations already described as special cases.

Jonscher has described the pattern of behaviour in Hill's examples and a variety of others as the 'universal law of dielectric response', at least for solids, and further concluded that it was necessary to 'seek a universal framework within which all these materials would find a common interpretation'. It is not clear from Jonscher's various presentations, however, whether the 'universality' refers both to limiting low- and high-frequency forms of equation (17) or only to the $(\omega\tau_0)^{-(1-n)}$ high-frequency dependence. In any case, we shall come to a very different conclusion from discussion of recent theoretical developments of correlation functions for specific molecular models in §6.

Although the empirical equation (17) resembles the Havriliak–Negami equation for suitable choices of the parameters, and the other empirical equations with more restricted choices, the predictions are not the same. The differences in quality of fit possible should be quite readily distinguishable for data of reasonable precision over a few decades of frequency in the dispersion region. Jonscher and Hill do not discuss such questions, and it is not clear to the writer what the answers are for the various examples. It is interesting, however, that Hill (1978) has reported that a more general expression of the form

$$\epsilon'' \sim \frac{(\omega\tau_0)^m}{[1 + (\omega\tau_0)^{2s}]^{(m+1-n)/2s}} \tag{18}$$

could by adjustment of the three parameters m, n, and s (all in the range 0 to 1) give 'remarkably good agreement with the experimentally determined plots'. As he points out, in this form the parameter s modifies the Jonscher behaviour only near the loss peak and by adjustment can give either the Cole–Cole or Fuoss–Kirkwood functions for $m = 1-n$.

Before going on to a discussion of theories which have been advanced to account for the various types of behaviour described in this and the preceding section, one type of deviation from the high-frequency power law behaviour which is not uncommon should be mentioned. This a change from a power law $(\omega\tau_0)^{-\beta}$ with β greater than 0.5 over a range of frequencies with $\omega\tau_0 > 1$ to a still more gradual variation with a new β' of order 0.5 for $\omega\tau_0 \gg 1$. Yoshihara and Work (1980) point this out for the amorphous polymer poly(4-chlorostyrene). Similar behaviour has been seen for viscous polar liquids such as alkyl halides, and comments by Böttcher and Bordewijk about several other systems seem to indicate similar behaviour for them. An obvious possible explanation is that a distinct faster relaxation process is becoming evident, and this was established for low-temperature relaxation of solid hydrogen bromide, for example. It is also possible, however, that it can be a limiting behaviour of the relaxation process which is not accounted for by the empirical descriptions. We return to this question in § §5 and 6.

5. Defect diffusion and fluctuation models

The fact that relaxation processes in so many substances show characteristic patterns of large deviations from simple exponential or Debye behaviour has led to increasing recognition that considerably better account needs to be taken of the roles of inter- and intramolecular forces in effecting charge displacements than is provided by simple diffusion or barrier models. The formal correlation function expressions given in §2 provide a very general prescription for doing this, but by design only in terms of effects of the many-body Liouville operator incorporating any and all forces. The basic necessity in this or any other formulation is of course to devise approximations which are not too drastic and make solutions possible.

If one grants that numerical solutions of even grossly simplified equations of motion are not feasible for the very long times, on a molecular scale, for loss of correlations, then stochastic or probabilistic representations of some kind must be introduced as at least a partial substitute for explicit molecular dynamics. We begin our discussion of a few of the numerous attempts to do this by considering two approaches which rely mainly on stochastically based concepts and give some useful insights into how major deviations from Debye behaviour can come about.

5.1. The Glarum defect-diffusion model

In a first approach, Glarum (1960) proposed his defect-diffusion model in which the decay of initial polarisation of dipoles results from two causes. The first, as in rotational diffusion or jumps between barrier sites, is assumed to give a rate of decay proportional to the polarisation at the time. On the presumption that such a mean field approximation takes inadequate account of fluctuations in the intermolecular forces, a somewhat *ad hoc* and certainly oversimplified assumption is made that such fluctuations, e.g., of energy or local free volume, are also operative which relax or randomise the orientations of a dipole completely. It is then further assumed that the times for this to occur are given by the probability of first arrival at a time t of any one of a distribution of such defects which diffuse through the sample as a one-dimensional random walk.

The correlation function $\Phi(t)$ for the model is then given by

$$\Phi(t) = \exp(-t/\tau_0)\,[1 - P(t)]$$

where $P(t)$ is the probability of arrival at time t of the nearest defect in a random distribution at $t = 0$ with average distance l. This is taken from diffusion theory to be given by

$$\dot{P}(t) = \frac{1}{\tau_d} [(\pi t/\tau_d)^{1/2} - \exp(t/\tau_d) \operatorname{erfc}(t/\tau_d)^{1/2}]$$

where $1/\tau_d = D/l^2$ is a diffusive rate constant and erfc is the complementary error function. Solution for the relaxation function $L[-d\Phi/dt]$ by Laplace transforms gives

$$L[-d\Phi/dt] = \frac{1}{1 + i\omega\tau_0} \left[1 + \frac{i\omega\tau_0}{1 + (\tau_d/\tau_0)^{1/2}(1 + i\omega\tau_0)^{1/2}}\right] \tag{18}$$

Depending on the ratio of the diffusive relaxation time to the dipole relaxation time, this gives relaxation functions changing from Debye behaviour for $\tau_d \gg \tau_0$ through increasing high-frequency deviations to the skewed-arc function with $\beta = 0.5$ for $\tau_d = \tau_0$, then a development of low-frequency deviations to give the circular-arc function with $1 - \alpha = 0.5$ in the limit of pure defect relaxation for $\tau_0 \gg \tau_d$. For intermediate values with $\tau_d > \tau_0$, the function resembles the skewed-arc function with values of $\beta > 0.5$ (for $\beta = 0.7$, $\tau_d/\tau_0 = 5.4$ for example). The principal difference is a high-frequency 'tail' varying as $(\omega\tau_0)^{-1/2}$ for $\omega\tau_0 \gg 1$, and this is one basis for the comments about such behaviour at the end of the preceding section. For $\tau_0 > \tau_d$, the low-frequency asymmetry resembles that of the Havriliak–Negami function for suitable values of $1 - \alpha$ and β.

The shortcomings of the defect-diffusion model are sufficiently obvious that it is clearly not to be taken literally. Rather, its interest lies in the fact that simulations of various power law dependences can be generated by diffusion-like processes with characteristic $t^{1/2}$ time dependence if these operate together with simple exponential relaxation processes. This is an example of the proposition that two seemingly quite different functions can predict very similar behaviour.

Glarum's model has been modified and extended in several ways: inclusion of relaxation by next nearest and all neighbouring defects, and defect-diffusion in three dimensions. These developments are reviewed in Böttcher and Bordewijk (1978).

5.2. The Anderson–Ullman fluctuation model

Anderson and Ullman (1967) have developed a model, stimulated by Glarum's approach, of the effects of fluctuating environment on molecular relaxation. The reorientation of a particular molecule is assumed to be determined by a transition probability or rate which is a function of, loosely, 'free volume' or other parameters whose fluctuations in time about a mean volume are governed by diffusion-like or Brownian motion in a force field restricting their amplitude. The picture they develop from numerical evaluations of the correlation and relaxation functions is that initially molecules with fast reorientation rates will do so in a more slowly changing environment, and hence at a variety of rates, but with increasing time the more slowly relaxing molecules will do so more nearly in an average environment, and hence at more nearly the same rates.

With plausible functional dependences of rate on magnitude of fluctuation, Anderson and Ullman were able to generate relaxation functions varying from Debye behaviour for rapid fluctuations of rate, to a skewed-arc function with $\beta = 0.6$ in one case, to symmetric functions of roughly circular-arc form with $1 - \alpha = 0.3$ for slow fluctuations over a considerable range of rates. The exact form of the results obtained depended on the fluctuation-rate relationship employed, but by the nature of the model it is impossible to generate a relaxation function with a more gradual frequency dependence at low frequencies than at high.

As emphasised by the authors, the dynamic properties of the environment and coupling of individual molecules to it are represented by fluctuations in a single parameter, and without any prescription from which its effect on rates of reorientation can be deduced rather than assumed. The model thus leaves much to be desired as a stochastic description of fluctuating force fields, but like Glarum's model, does give a picture of how they can produce loss of correlations resembling those observed in dielectric relaxation.

6. Many-body relaxation models

In this section, we consider models in which several different kinds of interactions of molecules, chain segments, or other entities lead to strong deviations from simple exponential relaxation.

6.1. *One-dimensional Ising model*

Motions of dipoles perpendicular to a linear chain connecting them have been modelled in several ways. A classical prototype is the one-dimensional Ising model, in which effects of adjacent nearest neighbour interactions have been worked out explicitly for specific assumptions about the dependence of transition probabilities for switching between two states of a chain element on the states of its two neighbours. Anderson (1970) has evaluated the resulting correlation functions for a dipole, or dipoles, attached to such a chain, using methods and results of Glauber (1963). Denoting the two states by one or zero, the assumed set of probabilities for reorientation was taken to be of the form

$$0\text{–}0\text{–}0 \xrightarrow[2K_1]{} 0\text{–}1\text{–}0;\; 1\text{–}0\text{–}1 \xrightarrow[2K_2]{} 1\text{–}1\text{–}1;\; 0\text{–}0\text{–}1 \xrightarrow[K_1+K_2]{} 0\text{–}1\text{–}1$$

with modified conditions for the end elements of a finite chain. The probabilities of the 2^N configurations for N elements are then coupled by a set of 2^N linear rate equations, which could in principle be solved for the distribution of relaxation times of the normal modes and the correlation function for the dielectrically active ones. Fortunately, the hopelessly unwieldy problem for large N can be reduced to manageable proportions by introducing N probabilities P_i of finding each chain element in one of its two states, which results in N coupled rate equations for the P_i.

Anderson obtained numerical solutions for various assumptions about the rates $R = K_2/K_1$, number of chain elements, number and spacing of dipoles associated with the elements, and end conditions. The derived correlation functions exhibited such behaviour as simple exponential decay for a single dipole $R = 1$, a skewed-arc for $R \gg 1$ and $R \ll 1$, and two distinct well separated relaxation regions for $R \gg 1$ and low chain end mobilities.

We mention here only a few significant points about the many results described by Anderson. One is that the coupling in the model which leads to broadened loss peaks for a *single* dipole is obviously from constraints of the chain dynamics, not from joint correlations of interacting dipoles. These correlations for more than one dipole on the chain can narrow, broaden, or otherwise modify the relaxation behaviour for only one, but again as a further effect of chain dynamics rather than dipole coupling energy. The overall relaxation time and frequency of maximum loss are insensitive to chain length, rather than increasing as would be expected for dipoles parallel to the chain backbone.

The results for the Ising model have qualitative features like those of real chain polymers, but as Anderson points out this hardly establishes the validity of the assumed dynamics to the exclusion of quite different models which can lead to similar results. The next examples of chain motion dynamical models illustrate this point, about which more will be said in a more general way later.

6.2. *The Shore and Zwanzig model*

Shore and Zwanzig (1975) assumed a series of elements rotating in planes perpendicular to a common axis. Each of these was assumed to undergo rotational diffusion as a model of interactions with the external surroundings, and also to be subject to torques proportional to the cosines of the differences in angles of rotation of the nearest neighbours. For strong coupling along the chain, the interaction energy U is then

$$U = -J \sum_i \cos(\theta_i - \theta_{i+1}) \simeq (J/2) \sum_i (\theta_i - \theta_{i+1})^2 + \text{constant}.$$

A forced diffusion equation including the effects of the corresponding neighbour torques of the model was then solved for chains with various numbers N of elements, relative strength $\beta J = J/kT$ of the neighbour interactions, and either open or closed chains (i.e., with periodic boundary conditions). The resulting distribution was then used to evaluate correlation functions for dipoles on various numbers of elements.

For a single dipole, Shore and Zwanzig find that the normalised correlation function $\exp(-Dt)$ with diffusion coefficient D for no coupling of elements is replaced quite generally by a time evolution through three regimes

$$\frac{\Phi(t)}{\Phi(0)} = \begin{cases} \exp(-Dt), & \beta JDt \ll 1 \\ \exp[-(Dt/\pi\beta J)\,1/2], & 1 \ll \beta JDt \ll N^2 \\ \exp(-Dt/N), & \beta JDt > N^2 \end{cases} \tag{19}$$

The 'free diffusion' with relaxation time $1/D$ at short times is thus soon replaced by a relaxation having the form of the Williams–Watts empirical function with $\beta_w = 0.5$, and finally a relaxation time for combined motion of all elements which is N times longer.

Numerical evaluation of the transforms to obtain $\epsilon^*(\omega)$ for a chain of 501 elements with a single dipole and $\beta J = 10$ gave a complex plane plot for ϵ^* which is closely fitted by a Havriliak–Negami function with $1 - \alpha = 0.83, \beta = 0.51$. For much shorter chains, $N < 50$, the deviations from Debye behaviour became considerably smaller and less confined to high frequencies ($\omega > \omega_m$). A rather similar behaviour is found when more dipoles are attached to the chain: for 101 elements with dipoles on every 10th element and $\beta J = 10$, the asymmetry reduced ϵ''_m by about 20% only from the Debye value 0.5 $(\epsilon_s - \epsilon_\infty)$, as compared to ϵ''_m about one-half the Debye value for the single dipole with $N = 501$ and the same βJ. For dipoles on every element, the relaxation is approximately fitted by the Cole–Davidson function with $\beta = 0.85$. One then has the result that the effect of joint correlations of dipoles is to *decrease* the deviation from Debye behaviour brought about by torques of neighbour orientations which are *not* of dipolar origin.

6.3. *Orientational dynamics of alkane chains*

Another kind of study of conformational dynamics in shorter chain molecules, such as alkanes with carbon–carbon bonds, which is receiving increasing attention is the use of Brownian dynamics computer simulations, in which the interaction with the surroundings is modelled by stochastic random forces chosen by Chandrasekhar's (1953) prescription with constraints of carbon–carbon bond orientations and steric hindrance with trans and gauche preferred orientations included. Using modified forms of computer programs developed by Fixman (1978), Glenn Evans has, with much encouragement and a little help from the writer, computed correlation functions of a terminal bond dipole from the model for several alkane chains using tetrahedral bond angles and a three-fold potential barrier $U(q) = (Q/2)(1 - \cos 3q)$ for the dihedral angle q. For the three-bond case of butane, the shortest alkane with steric hindrance, Evans has found departures from exponential decay which are approximately of Cole–Davidson form with $\beta = 0.856$ for a barrier height $Q = 5kT$ (~ 3 kcal mol^{-1} at room temperature), while rather smaller deviations of the same form are found for rigid and freely rotating chains. For an eight-bond chain, nonane, larger deviations are found, corresponding to $\beta = 0.63$ for the $5kT$ barrier, with the characteristic time for this barrier about four times that for free rotation and 0.4 that for the rigid chain.

These results are preliminary and like other numerical simulations do not of themselves give much insight into the physics of the interactions which may rapidly overcome the initial independence of chain motions. They also need some remodelling before making any close comparison with experimental results on alkyl halides as classic examples of Cole–Davidson behaviour, but the patterns of behaviour emerging from serious attempts to treat realistic chain conformations are encouraging.

6.4. *Fluctuating barriers of electrostatic origin*

Brot and Darmon (1970), in a very interesting paper which has not received the attention it deserves, used Monte Carlo methods to calculate both the static and time-dependent correlation functions of 1,2,3-trichlorotrimethylbenzene (TCTMB). This and two other hexasubstituted benzenes, 1,2-dichlorotetramethylbenzene (DCTMB) and pentachlorotoluene (PCT), have considerable orientational freedom in their crystalline solid phases. For TCTMB and DCTMB, with relatively large dipole moments (3.2 and 2.8 debye), orientational orderings set in gradually over temperature ranges of 100–150 K, and the ϵ^* data were fitted by circular-arc functions with increasing values of α at lower temperature. For PCT with dipole moment 1.55 debye the relaxation changed from Debye form at 294 K to skewed-arc functions with β decreasing to 0.55 at 186 K with the $1/T$ dependence of static permittivity for a disordered solid.

In the solid phases, the molecules are arranged in parallel layers with six-fold rotational disorder of the benzene rings in the layers inferred from x-ray studies. The orientational energy and fluctuations of rotational orientation barriers of TCTMB were represented by electrostatic interaction energies of a twelve-point charge model of the molecules which changed the potential well depths for a given molecule according to the relative orientations of the dipoles of the molecule and its neighbours. The static correlation function (Kirkwood g factor), calculated by Monte Carlo trials for 240 molecules, decreased roughly as observed experimentally from 0.5 at 400 K to less than 0.1 at 100 K. The time-dependent correlation functions calculated with a minimum barrier height of 2 kcal mol^{-1} was of Debye form within computational error at 300 K, but at 186 K the calculated function was fitted by a circular-arc with $\alpha = 0.28$ as compared to the observed $\alpha = 0.39$ at this temperature.

The results of Brot and Darmon are thus a very satisfactory demonstration that variations in the orientation-dependent activation energies of the any-jump model used leads to a dipole relaxation function with temperature-dependent form similar to that found experimentally.

6.5. *Stochastic trapping charge transport*

It is well known that charge transport in many disordered or amorphous solids shows considerable deviations from Debye-like behaviour. Experimentally, such systems as amorphous arsenic selenide, selenium, and polymers with high concentrations of dopants exhibit photoconduction currents of carriers, injected by a light flash of short duration, with a time dependence of the form $t^{-\alpha}$ at short times changing to an asymptotic variation as $t^{-(2-\alpha)}$ at long times, with α in the range zero to one. The exponents used here differ from the usual ones in the literature for such systems by the change from α to $1-\alpha$ to emphasise that the behaviour has the form of the transient discharge current for the empirical circular-arc relaxation function (Cole and Cole 1942).

Several models of the elementary processes have been proposed. One of particular interest for the present discussion is the stochastic transport model developed by Scher and associates, as described in Scher and Montroll (1975) and other references cited there. Here we indicate very briefly only the basic features of the model which lead to the circular-arc behaviour. As Scher and Montroll present the approach, for processes of charge trapping between cells containing one or more sites for the carriers, the probability $P(l, t)$ that a carrier is in cell l at time t is described in conventional charge transport theory by a master or rate equation of the form

$$\frac{\mathrm{d}P(l,t)}{\mathrm{d}t} = \lambda \sum\nolimits_{l'} [p(l-l')P(l',t) - p(l'-l)P(l,t)],$$

with the two sets of terms on the right giving the net flow from and to other cells l', with transition probability p. In the generalisation of the Montroll–Weiss (Montroll and Weiss 1965) continuous-time random walk model used, the rate constant λ is replaced by a relaxation function $\dot{\Phi}(t)$ to give a non-Markoffian master equation

$$\frac{\mathrm{d}P(l,t)}{\mathrm{d}t} = \int_0^t \mathrm{d}x\,\Phi(t-x) \sum_{l'} [p(l-l')P(l',t) - p(l'-l)P(l,t)], \qquad (20)$$

Usually only transitions between nearest neighbours l and l' are considered and $\Phi(t)$ is related to a trapping time distribution function $\psi(t)$ giving the probability that a carrier will leave for its next cell at time t after arrival, by the integro-differential equation

$$\frac{d\psi}{dt} = \psi(0)\,\delta(t) + \Phi(t) - \int_0^t dx\,\Phi(t-x)\,\psi(x).$$

For $\psi(t)$ decaying at long times as $1/t^{2-\alpha}\,T(\alpha)$, this results in the observed transient currents varying as $t^{-\alpha}$ at short times, followed by a transition, as times for passage of carriers to an absorbing boundary are reached, to the $t^{-(2-\alpha)}$ variation. Contrary to some suppositions, the trapping function $\psi(t)$ which accomplishes this is not purely empirical, as Scher and Lax (1973) obtained a function of similar form by averaging over a spectrum of transition rates between sites determined by wavefunction overlap with the coefficient α determined by density of sites and effective overlap distance.

An alternative model, discussed by Lakin *et al* (1977) is that the charge carriers move through finite numbers and distinct types of traps characterised by different lifetimes and release times, and Noolandi (1978) has shown that the multiple trapping equations for the model can be equivalent to the Scher–Montroll master equation in its continuum limit. As a result, the fractional power law current decay functions can be approximated over considerable time ranges by discrete distributions of traps with varying depths producing the counterpart of a distribution of relaxation times. The necessary number and range of characteristic times was found to be small for usual ranges of time, but this could presumably be tested further by adequate measurements over several decades of time in the appropriate range.

6.6. The model of correlated states and configurational fluctuations

A many-body theory of dielectric relaxation, primarily in solids, has been developed by Jonscher and co-workers in a series of papers. As their viewpoints and formulations are presented in the paper by Jonscher in this volume, only a brief summary of some of their basic ideas is given here for comparison with the models for specific systems which we have discussed in this section.

The development by Dissado and Hill (1979) is based on a model which considers the time evolution of macroscopic average moments of a system in which transitions between different configurational states separated by an energy barrier take place by three distinct processes:

(a) conventional thermally activated passages over the barrier;
(b) local 'flip' processes in which the total polarisation changes by 'configurational tunnelling' to produce local changes of dipole or spin orientations;
(c) local 'flip-flop' processes involving two opposed configurational changes, which acting alone leave the polarisation unchanged but modify the energy distributions in the states.

The process (ii) is identified as the origin of the short-time power law behaviour by a time-dependent quantum mechanical perturbation theory calculation of transitions from excited initial states averaged over a range of energy differences in the final state. This gives a t^{-n} dependence for the assumed spread of available states with n a measure of correlations between such transitions determined by the available excitation energy and coupling of the states. This model of short-time behaviour together with exponential decay from activated barrier crossing gives a behaviour in the absence of other processes very like that of the Cole–Davidson and Williams–Watts empirical functions for appropriate values of n and β or β_W. The coupled 'flip-flop' processes in the model producing local rearrangements and fluctuations of macroscopic moments, in times estimated to be of order 10^{-8} to 10^{-10} s, lead to the third factor coming into play at long times with an asymptotic variation of the relaxation rate as $t^{-(1+m)}$ and an interpretation of m similar to that for n.

The model thus attributes behaviour of the form described by the Havriliak–Negami and Hill empirical formulae to the sequence of 'flip' processes dominating the short-time behaviour, quasi Debye-like behaviour only at times of order $1/\omega_p$ where ω_p is the frequency of the loss peak, and correlate 'flip-flop' transitions as the dominant relaxation mechanism. One consequence of the model is that relaxation of the skewed-arc form would be expected only at frequencies above 10^7 Hz or for perfect correlation of the flip-flop processes ($m = 1$). While it is true that such relaxation is found in liquids and disordered rotator phases of solids at high frequencies, and is in agreement with the first limitation, there are numerous examples of viscous liquids in which such relaxations have been found at much lower frequencies (<1 Hz in some cases), requiring perfect correlation of 'flip-flop' processes if these are of general occurrence and considered essential.

Some of the basic features of the Dissado and Hill model which lead to non-exponential relaxation have no apparent resemblance to the interaction mechanisms incorporated in the more specific relaxation models which also lead to non-exponential behaviour. Dissado and Hill invoke dipole–dipole interactions which are either not involved in the chain models or, if they are, can lessen the deviations from Debye behaviour, and their identification of an intermediate Debye-like region for times of order $1/\omega_p$ has no evident counterpart in the other models. In the concluding section we consider other questions, such as the 'universal' applicability of the correlated state approach in preference to a number of models which are more explicitly formulated for specific types of system and many-body interactions.

7. Conclusions

A considerable number and variety of interpretations of dielectric relaxation and charge transport processes developed in recent years are making the subject a lively and controversial one. There appears to be fairly general agreement that large deviations from exponential behaviour at both short and long times often result from many-body effects rather than as sums of physically distinct simpler processes. Jonscher (1975a, b) has emphasised similarities of time and frequency dependences found in a wide variety of systems: liquid, solid, or glassy; amorphous or crystalline; semiconductors or molecular. Readers will have to decide for themselves whether and how literally this is to be described as 'universal behaviour.' The writer would be content to suggest that general similarities and individual differences, observable with limited resolution, are both important.

Jonscher has argued that the universality, or at least remarkable similarity, leaves no room for escape from the conclusion that there must be a single kind of mechanism and that all previously proposed mechanisms for particular systems must be rejected because they are too special to account for the generality of behaviour. The writer had somewhat the same opinion many years ago, as put forward in early papers with K S Cole (Cole and Cole 1941, 1942). Since then, however, the development of increasingly plausible and realistic molecular theories for particular kinds of systems with quite different models of intermolecular forces, but strikingly similar types of prediction about frequency and time dependences, has led me to think otherwise. The models discussed in the preceding section seemed particularly good examples of the diversity possible; others appropriate to other situations could have been added. Anderson (1970), however, has pointed out the dangers in all such discussion by his statements that 'The fact that a given model fits certain experimental data provides no real assurance that the physical relationships going into the model are found in nature' and that 'Often models exphasising rather different physical ideas fit the same data equally well. Witness the many models consistent with purely exponential relaxation.' Similar comments by Shore and Zwanzig (1975) are equally relevant.

The fractional power law dependences so widely used to represent experimental results present the intriguing problem of whether they should be generated by a proper analytical treatment. Dissado and Hill did obtain such asymptotic results from their model, as did Lewis (1977) for plausible assumptions about energy states involved in charge transport in amorphous solids, and other examples could be added, notably for $t^{-1/2}$ time dependences from diffusion models. At the same time, however, one could cite a variety of analytical functions and sums over normal modes which have no variable fractional powers appearing in them but behave remarkably like fractional power laws over considerable ranges of time or frequency for suitable values of other parameters.

It appears to the writer that the search for and development of better models will continue, aided greatly by more sophisticated and powerful numerical methods. Dielectric relaxation studies as a probe of the real underlying dynamics have their limitations as discussed in §2 and elsewhere, as do all macroscopic relaxation measurements, but do have great advantages in precision and in the range of frequencies (10^{11} to 10^{-3} Hz at least) which can be covered with existing methods. With this range one can, with varying degrees of difficulty to be sure, hope to define virtually complete relaxation functions at any fixed temperature in a reasonable range without resorting to time–temperature superpositions of data for fixed limited ranges of frequency. These work well if the *form* of the frequency dependence is independent of temperature, but can give a distorted representation if there is a temperature dependence, and are to be avoided if possible. Several examples of such dependence have been mentioned and others could be added; force fitting such data with temperature-independent parameters both distorts and obscures the nature of the real behaviour. It is also important to consider both the temperature dependence and magnitude of the static permittivity, as this is determined by the equilibrium value of the dipole correlation function. Moreover, merely determining the sign of $\partial\epsilon_s/\partial T$ establishes whether applying an electric field increases or decreases the entropy of the system by use of the Maxwell relation $\partial\epsilon_s/\partial T = K(\partial S/\partial E^2)$ with K a positive constant set by the choice of units.

Finally, we should mention the importance of obtaining and combining evidence from other kinds of relaxation measurements with dielectric response results to obtain more definitive information about molecular dynamics than is possible from any single kind of relaxation measurement. The point has been properly emphasised by Graham Williams (1978), and results of combined measurements of dielectric and Kerr effect relaxation measurements by Beevers *et al* (1976) on viscous dipolar solutions show, for example, how one can distinguish between small- and large-angle orientational jumps. Molecular correlation function expressions have been developed for a variety of linear mechanical and magnetic, as well as electrical, effects and the formal results both show what aspects of molecular dynamics are involved in the different cases and provide useful bases for comparison of predictions from specific models.

With the increasing availability of strong perturbing fields from laser and other technologies, there is both a need for and value receivable from developing nonlinear response theory treatments to determine what information about correlation functions can be deduced from such nonlinear effects as transient electric birefringences (dynamic Kerr effect) and nonlinear electric permittivity in strong fields. The difficulties in nonlinear response theory have been discussed in general terms by Zubarev (1974) and for the Kerr effect problem by Cole (1977). The writer has also been able to obtain expressions for quadratic transient Kerr effect responses to the application and removal of a polarising field. The results to be presented elsewhere (Cole (1981), and in preparation) show that for dipolar systems one can obtain information about correlations of both $P_1[\cos\theta(t)] = \cos\theta(t)$ and $P_2[\cos\theta(t)] = (1/2)(3\cos^2\theta(t) - 1)$. There are moreover, encouraging indications as to how one may be able to make progress in obtaining formal expressions for higher-order nonlinear effects with reasonable and nonspecific assumptions about the molecular dynamics. Availability of such results for dielectric and other relaxation behaviour in strong fields should provide valuable tools for analysis of new effects obtained by using strong perturbing fields as probes of dynamical response.

References

Anderson J E 1970 *J. Chem. Phys.* **52** 2821

Anderson J E and Ullman R 1967 *J. Chem. Phys.* **47** 2178

Beevers M S, Crossley J, Garrington D C and Williams G 1976 *J. Chem. Soc. Faraday II* 72 1482

Böttcher C J F and Bordewijk P 1978 *Theory of Electric Polarization* vol II (Amsterdam: Elsevier)

Brot C and Darmon I 1970 *J. Chem. Phys.* **53** 2271

Chandrasekhar J 1943 *Rev. Mod. Phys.* **15** 1

Cole K S 1928 *J. Gen. Physiol.* **12** 29

Cole K S and Cole R H 1941 *J. Chem. Phys.* **9** 341

— 1942 *J. Chem. Phys.* **10** 98

Cole R H 1955 *J. Chem. Phys.* **23** 493

— 1965 *J. Chem. Phys.* **42** 637

— 1977 *Ann. N. Y. Acad. Sci.* **303** 59

— 1981 to appear in *Biological Effects of Non-Ionizing Radiation,* ed. K H Illinger, ACS Symposium Series (Washington, DC: American Chemical Society)

Colonomos P and Gordon R G 1979 *J. Chem. Phys.* **71** 1159

Davidson D W and Cole R H 1950 *J. Chem. Phys.* **18** 1950

Debye P 1929 *Polar Molecules* (New York: Chemical Catalog Co,)

Dissado L A and Hill R M 1979 *Nature* **279** 685

Fixman M 1978 *J. Chem. Phys.* **69** 1527, 1538

Fuoss R M and Kirkwood J G 1941 *J. Am. Chem. Soc.* **63** 385

Glarum S H 1960 *J. Chem. Phys.* **33** 639

Glauber R J 1963 *J. Math. Phys.* **4** 294

Groenewegen P P M and Cole R H 1967 *J. Chem. Phys.* **46** 1060

Havriliak S and Negami S 1966 *J. Polym. Sci.* **C14** 99

— 1967 *Polymer* 8 161

Hill R M 1978 *Nature* **275** 96
Hoffman J D 1955 *J. Chem. Phys.* **23** 1331
Jonscher A K 1975a *Colloid and Polymer Sci.* 253, 231
— 1975b *Nature* **256** 566
Kirkwood J G and Fuoss R M 1941 *J. Chem. Phys.* **9** 329
Kubo R 1957 *J. Phys. Soc. Japan* **6** 570
Lakin W D, Marks L and Noolandi J 1977 *Phys. Rev.* **B15** 5834
Lauritzen J I 1958 *J. Chem. Phys.* **218** 118
Lewis T J 1977 *Dielectric and Related Molecular Processes* vol 3 (London: The Chemical Society) p187
Mashimo S 1976 *Macromolecules* **9** 91
Meakins R and Mulley J 1954 *J. Chem. Phys.* **21** 1934
Montroll E and Weiss G H 1965 *J. Math. Phys.* **6** 169
Noolandi J 1978 *Phys. Rev.* **B16** 4474
Patey G S 1977 *Mol. Phys.* **34** 427
Scher H and Lax M 1973 *Phys. Rev.* **B7** 4491, 4502
Scher H and Montroll E 1975 *Phys. Rev.* **B12** 2455
Shore J E and Zwanzig R 1975 *J. Chem. Phys.* **63** 5445
Toupin R A and Lax M 1957 *J. Chem. Phys.* **27** 458
Van Vleck J H and Weisskopf V 1945 *Rev. Mod. Phys.* **17** 227
Wertheim M S 1971 *J. Chem. Phys.* **55** 4291
Williams G 1978 *Chem. Soc. Rev.* **7** 89
Williams G and Watts D C 1970 *Trans. Faraday Soc.* **66** 80
Yoshihara M and Work R N 1980 *J. Chem. Phys.* **72** 5909
Zubarev D N 1974 *Non-equilibrium Statistical Thermodynamics* (New York and London: Plenum)
Zwanzig R 1963 *J. Chem. Phys.* **38** 2766

Inst. Phys. Conf. Ser. No. 58
Invited paper presented at Physics of Dielectric Solids, 8–11 September 1980, Canterbury

A many-body universal approach to dielectric relaxation in solids

A K Jonscher
Chelsea College, University of London, Pulton Place, London SW6 5PR

Abstract. The universality of the dielectric response of solids regardless of their detailed physical and chemical properties is based on well established experimental evidence. At times longer than, say, 10^{-10} s and extending to 10^5 s or more, this may be defined as the power law dependence of the depolarisation current

$$i(t) \propto t^{-s} \quad \text{with} \quad 0 < s < 1 \quad \text{or} \quad 1 < s < 2 \tag{i}$$

the exponent s taking two different values falling either in the first or in both ranges, in different intervals of time. In dipolar materials the resulting frequency-domain response gives a loss peak with power law branches below and above a characteristic loss peak frequency ω_p. In charge-carrier-dominated systems the high-frequency branch is similar, while the low-frequency branch shows a strong dispersion of both the real and imaginary components of susceptibility. A limiting form of behaviour is the frequency-independent loss, for which $s \to 1$, found in many materials as a background to all other loss processes. A new theoretical model is described which is based on the concept of many-body interactions leading to an infrared divergence-like response and explaining the totality of the observed forms of behaviour summarised by equation (i). This model considers, in addition to thermally assisted transitions of dipoles or charges, two different classes of configurational tunnelling which do not involve any thermal exchange with the heat bath. The exceptional sensitivity of dielectric response to many-body interactions is being used to develop a theoretical picture which has application much wider than the interpretation of dielectric processes.

1. Introduction

The frequency dependence of the dielectric permittivity $\tilde{\epsilon}(\omega) = \epsilon'(\omega) - i\epsilon''(\omega)$ has been studied in the last century in a very wide range of materials, over a frequency range approaching 15 decades and through a wide range of temperatures extending from the highest compatible with the stability of the materials in question and down to below 1 K (Jonscher 1980a). The study of the dielectric response becomes much easier if the various processes are sufficiently well separated in frequency so that they do not overlap significantly, which is often the case. The dielectric behaviour in particular regions of the frequency spectrum is then best represented in terms of the dielectric susceptibility

$$\tilde{\chi}(\omega) = \chi'(\omega) - i\chi''(\omega) = \tilde{\epsilon}(\omega) - \epsilon_\infty \tag{1}$$

where ϵ_∞ is the limit to which ϵ' tends at frequencies which are sufficiently high for the particular polarisation mechanism in question to show negligible loss and dispersion.

The most conspicuous features of the dielectric response are the loss peaks which normally occur at thermally activated frequencies $\omega_p(T)$ and whose shape departs in varying degrees from the ideal but singular Debye response:

$$\tilde{\chi}(\omega) \propto (1 + i\omega/\omega_p)^{-1} \tag{2}$$

Since we are dealing with the frequency dependence of a complex parameter, there are many ways of representing the experimental information, e.g. the complex susceptibility plot, i.e. the Cole–Cole representation, and the linear plot of $\chi'(\omega)$ and $\chi''(\omega)$ against $\log \omega$. These representations are very useful for study of the neighbourhood of loss peaks which are emphasised in these graphs but the details of the response away from the loss peak are thereby lost to a large extent.

The Fourier transform of equation (2) gives the time-domain response of the depolarisation current under the action of a step-function field:

$$i(t) \propto \exp(-\omega_p t) \tag{3}$$

and it may be noted that any physical mechanism leading to an exponential decay of polarisation necessarily gives a Debye response.

Some branches of the subject, notably those specialising in the electrochemical aspects of the response of ionic conductors, prefer to plot the complex impedance representation which is defined in terms of the complex capacitance $\tilde{C}(\omega)$ as $\tilde{Z}(\omega) = 1/[i\omega\tilde{C}(\omega)]$. This representation is especially useful when the material under investigation is not homogeneous, e.g. due to the presence of a series barrier near one of the electrodes. In this case a meaningful interpretation of the dielectric permittivity or susceptibility of the presumed 'bulk' material is only possible after subtraction of the relevant 'barrier' impedance (Jonscher 1978a).

Some schools, especially those dealing with the relatively more conducting materials, e.g. amorphous electronic semiconductors, prefer to plot the 'alternating current' (AC) conductivity

$$\sigma(\omega) = \sigma_0 + \sigma'(\omega) = \sigma_0 + \omega\,\epsilon''(\omega) \tag{4}$$

where σ_0 is the zero-frequency limit, or 'true DC' conductivity, and $\sigma'(\omega)$ is the 'true AC' conductivity. This presentation avoids the singularity at $\omega = 0$ arising from the term σ_0/ω in the dielectric loss representation.

The first purpose of this review is to give a systematic classification of the complete range of dielectric material responses in a wide range of frequencies below the terahertz range, thus excluding inertial, phonon and quantum effects. We shall arrive at the conclusion that the behaviour in this 'viscous' regime presents a remarkable uniformity or universality pointing strongly to the dominance of a common mechanism. The universality may be described in terms of the depolarisation current under step-function field, having the form:

$$i(t) \propto t^{-s} \tag{5}$$

in which the exponent s takes two different values at 'short' and 'long' times. In dipolar systems these values fall in the ranges $0 < s < 1$ and $1 < s < 2$. In systems dominated by hopping ionic or electronic charge carriers, s falls in the single range $0 < s < 1$, the 'short' time behaviour corresponding to a value closer to unity, as in many dipolar systems, while at 'long' times s lies closer to zero. These relations are shown schematically in figure 1.

The second purpose of this review is the presentation of a new many-body theory of dielectric response which is capable of explaining the observed behaviour in a much more unified fashion than the diverse interpretations currently accepted. In the interest of brevity, we shall omit detailed references to literature data, since these may be found in recent reviews (Jonscher 1977a, 1980 a, b). The same applies to the theoretical references which may be found in addition in Dissado and Hill (1979, 1980).

Our broad classification can most conveniently be established in a representation of $\log \chi'$ and $\log \chi''$ plotted against $\log \omega$, since this preserves the information far away from the loss peak and brings out the prevailing power law relationships. This advantage more than offsets the resulting loss of sensitivity in the peak region itself.

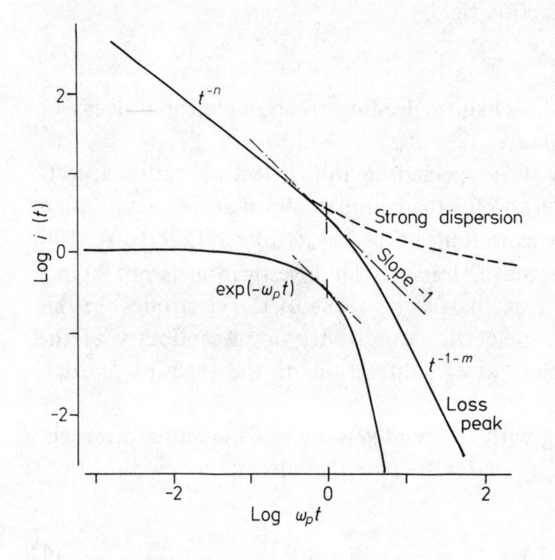

Figure 1. The logarithmic plot of the depolarisation current $i(t)$ for three cases. The lower curve represents the exponential decay corresponding to the Debye response. The upper curve shows the common t^{-n} relation which applies to both dipolar and charge-carrier systems in the universal response at 'short' times. The behaviour of charge-carrier and dipolar systems differs at 'long' times, as indicated. The time scale is normalised to the loss peak frequency ω_p, or the corresponding critical frequency for strong low-frequency dispersion ω_c.

The dielectric susceptibility $\tilde{\chi}(\omega)$ represents the response of the material medium and it may be due to several independent mechanisms which may overlap in any given frequency or time range. In singling out a prevailing 'universal' trend we may have to bear this clearly in mind. There are many examples of behaviour which correspond clearly to a single type of response in a range of between four and eight decades of frequency, showing that there is a single mechanism in operation. Equally, however, there are many examples of departures from an 'elementary' universal trend in any given range and one should not conclude immediately that the universal response does not exist at all. We should use the simple cases as establishing the existence of the universal behaviour and we should then see to what extent is it possible to regard the departures as the result of superposition of two or more elementary trends. The starting point for our present many-body approach is the existence of many well attested examples of the universal power law covering many decades of frequency, since these cannot be reconciled with any other simple theoretical model.

The point is sometimes made that the universality as defined in equation (5) cannot be valid in the entire time or the corresponding entire frequency range from zero to infinity, and that it is therefore somehow 'unphysical'. The reply to this is that very few physical laws are expected to be valid rigorously in the entire range of their respective variables and, in particular, the exponential Debye law cannot be valid in the inertial regime. Another criticism is that the solution of the theoretical model leading to a physically meaningful result should represent the solution of a simple differential equation. The new theoretical model can, in fact, be cast in the form of a differential equation (Halperin *et al* 1972) or alternatively an integral equation, but neither is simple and they must be solved self-consistently (L A Dissado, private communication). Finally, we mention the criticism that 'anything can be proved in the log–log representation'. There is no doubt, however, whether one is plotting a power law or an exponential relation and neither is there any doubt what are the values of the logarithmic slopes.

Shortage of space prevents us from quoting extensive examples of dielectric responses across the entire spectrum of materials and temperatures; the interested reader is referred to the reviews already mentioned, especially Jonscher (1980a) for more detailed treatment and for source references.

2. Review of experimental evidence

2.1. Dipolar materials

Figure 2 shows a compilation of dielectric loss data for several dipolar materials, ranging from peaks that are only slightly broader than Debye to shapes that are virtually unrecognisable in terms of the ideal behaviour. These examples refer to systems in which the *shape* of the loss curve is relatively independent of temperature. There are likewise many examples where this is not the case and the loss peaks become narrower as the temperature increases – this may be seen in some of the examples shown in figure 3. The dielectric peaks may be represented by the empirical expression (Jonscher 1975a):

$$1/\chi''(\omega) = (\omega/\omega_p)^{-m} + (\omega/\omega_p)^{1-n} \tag{6}$$

with both the exponents m and $1 - n$ falling in the range (0, 1). The Fourier transform of this expression may be represented by the approximate law for the time dependence of the depolarisation current:

$$i(t) \propto t^{-n} \qquad \text{for} \qquad t \ll 1/\omega_p \tag{7}$$

$$i(t) \propto t^{-1-m} \qquad \text{for} \qquad t \gg 1/\omega_p \tag{8}$$

which is the law shown in figure 1. In terms of equation (5) we have set $s = n$ and $s = 1 + m$, respectively. Equation (7) is the well known Curie–von Schweidler law, recognised for over seventy years as representing the widely observed behaviour of real dielectrics, as distinct from the idealised exponential behaviour expected of Debye processes. We note that $1/\omega_p$ corresponds approximately to the time at which $i(t) \propto t^{-1}$.

With reference to figures 2 and 3, we note that while there are many examples of the power laws shown in equation (6), there are also clear examples where the power laws are perturbed by the onset of some other law which in some cases may be recognised as a frequency-independent loss about which we shall say more later.

Although examples of near-Debye behaviour may be found in solids, where a careful analysis can be carried out on the basis of accurate measurements, significant departures

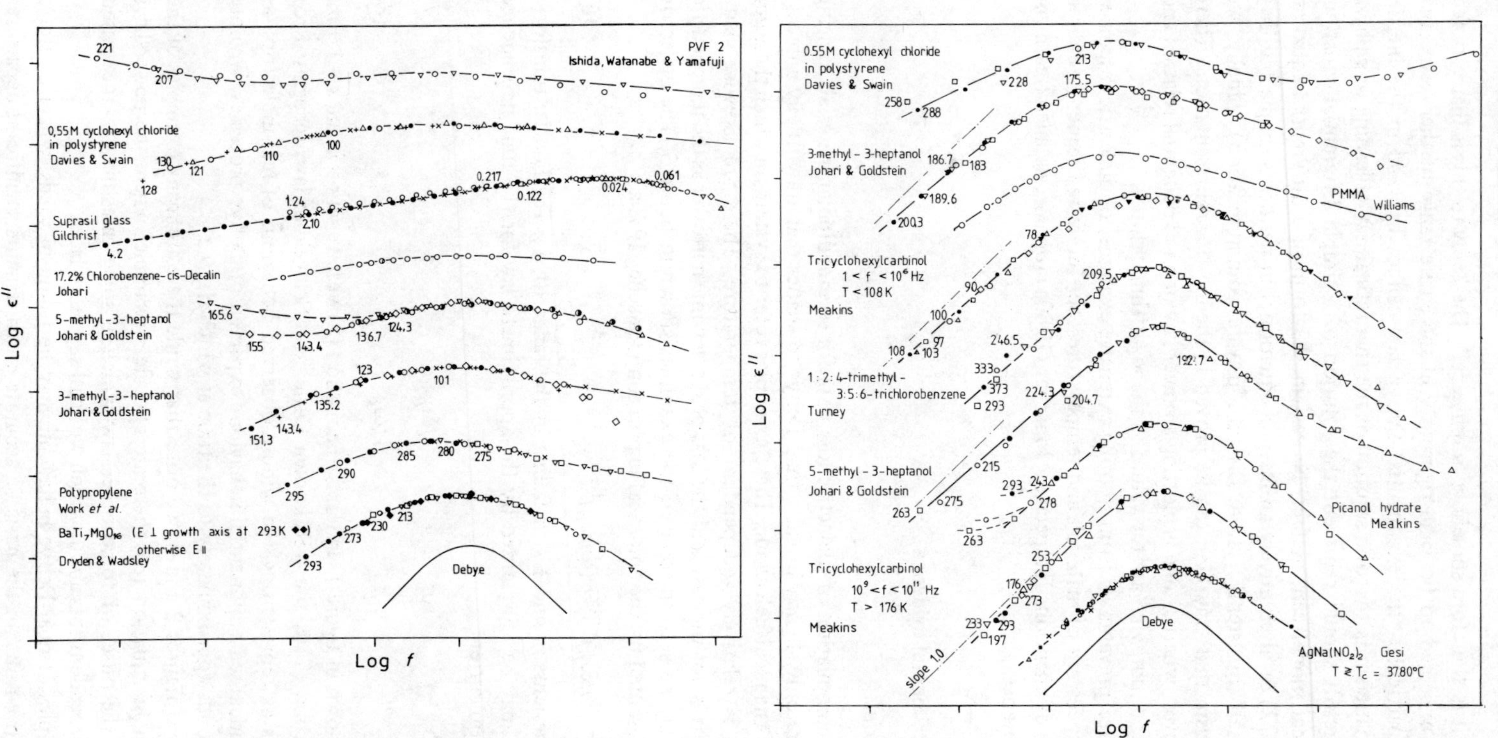

Figure 2. Two compilations of dielectric loss data for a range of dipolar materials plotted logarithmically, normalised for different temperatures and displaced both vertically and horizontally for clarity. The temperatures are indicated at the corresponding lowest frequency points. The left-hand set corresponds to the broader lower-temperature peaks, the right-hand set to the narrower higher-temperature peaks. Slopes corresponding to the limit $m = 1$ are indicated and so are the Debye responses. The sources of data are indicated. The logarithmic plots and the normalisations were performed by R M Hill who kindly supplied the individual diagrams.

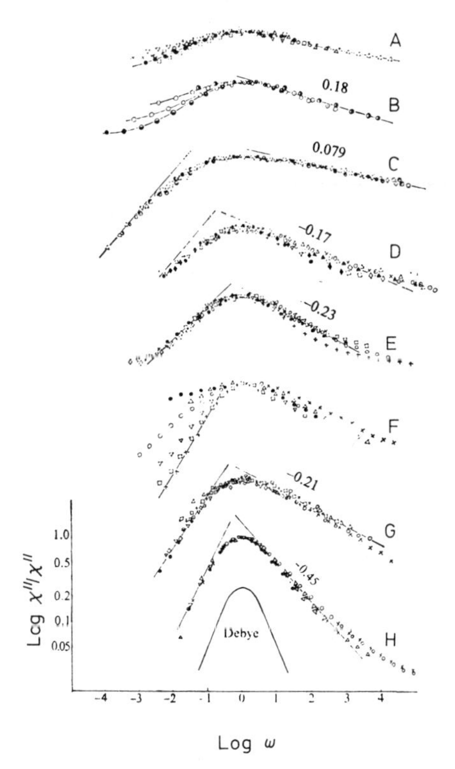

Figure 3. The compilation of dielectric data for a range of polymeric solids, plotted as log χ'' against log ω. Different symbols correspond to different temperatures, the data having been normalised by lateral displacement. The frequency axis is relative and the data have been displaced vertically for clarity. Note different logarithmic scales emphasising the sharpness of the peaks. (A) Polychloroprene – low-temperature data, (B) PCTFE, (C) polyethylene terephthalate, (D) polymethyl methacrylate, (E) polychloroprene high temperature data, (F) polyethyl methacrylate, (G) polydian carbonate, (H) polyvinyl acetate. (See Jonscher 1975a.)

can always be shown to exist with respect to the ideal equation (2). This requires that for $\omega \gg \omega_p$ we should have:

$$\chi''(\omega) \propto \omega^{-1} \qquad \text{and} \qquad \chi'(\omega) \propto \omega^{-2} \tag{9}$$

while the high-frequency limit of equation (6) and the Fourier transform of equation (7), together with the Kramers–Kronig transformations, give

$$\chi''(\omega) = \cot(n\pi/2)\,\chi'(\omega) \propto \omega^{n-1} \tag{10}$$

and this is valid however close n may be to zero, showing that the Debye limit represents a *singularity* in the family of universal relationships given by equation (6), and not merely a limit to which the system tends as the exponents $m = 1 - n \to 1$. This is a very sensitive test of the applicability of the Debye relations, as shown in figure 4 which gives an analysis of the data for single crystal CaF_2 doped with Er to 1 part in 10^4, which corresponds to a mean distance between dipoles equal to 22 lattice spacings (Fontanella *et al* 1978, Jonscher 1980c).

A wide ranging survey carried out by Hill (1978, 1980) has shown that the exponents m and $1 - n$ are *independent* of one another, so that any theoretical interpretation of the behaviour of dielectrics must allow the existence of two separate physical mechanisms. In terms of figure 1, these would correspond to two independent branches which represent *consecutive* processes with $s < 1$ and $s > 1$, respectively, giving rise to the appearance of a loss peak in the frequency domain (Jonscher 1975b). Hill's data are shown in figure 5

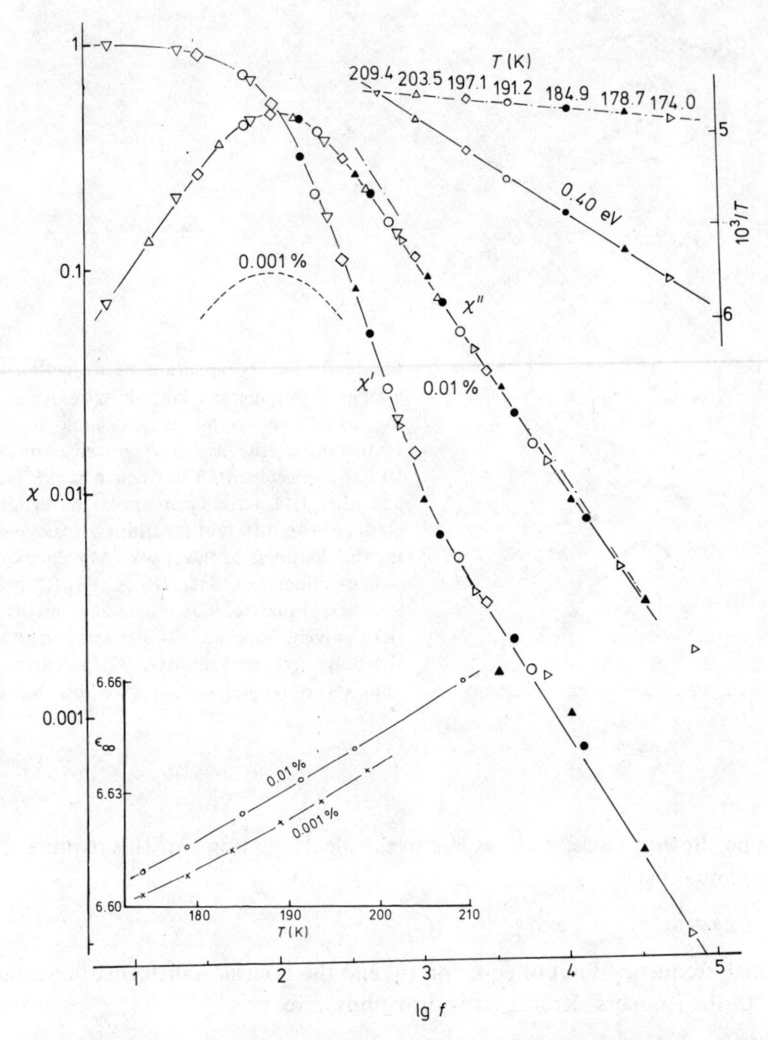

Figure 4. The real and imaginary components of the susceptibility of a single crystal of CaF_2 doped with 0.01% of Er, normalised by lateral shifting for a range of temperatures. The locus of the translation point is plotted at the top and is used to obtain the activation energy of 0.40 eV. The change of slope of the real part $\chi'(\omega)$ is clearly visible, showing a departure from the ideal Debye response, despite the near-Debye shape of the loss peak itself. The dotted line marks the corresponding position of the loss peak of a sample doped with 0.001% Er. The inset in the lower part shows the temperature dependence of ϵ_∞ for both samples. From Jonscher (1980c).

as a plot of the exponents m against $1 - n$ for 100 different materials. It is clear that the two exponents are independent of one another and that none of the empirical one-parameter representations, such as Cole–Cole, Cole–Davidson or Williams–Watts can adequately correspond to more than a small fraction of the data. We note, in particular, that some of the data refer to acoustic absorption, others to mechanical moduli, both of which obey very similar relations.

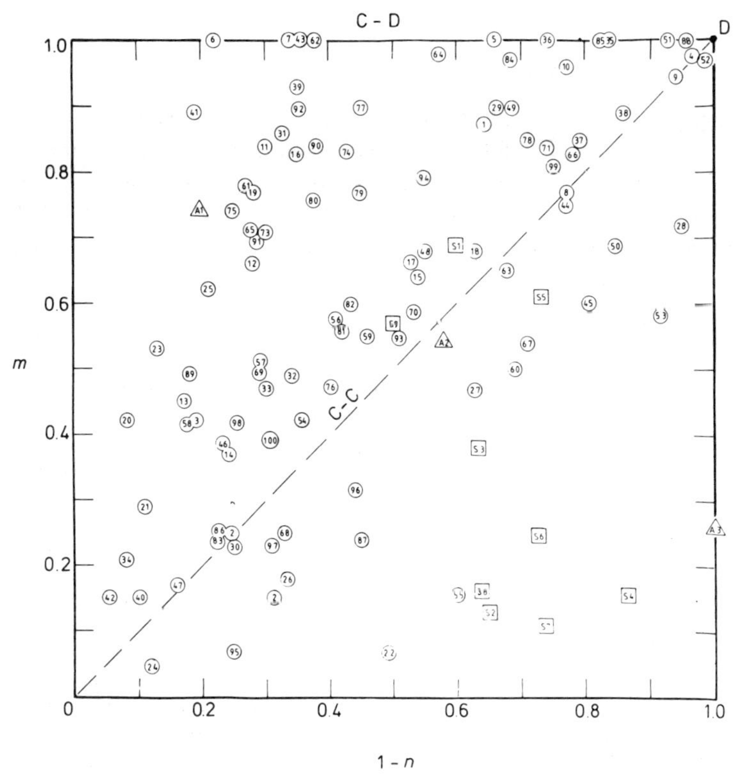

Figure 5. The representation of the dielectric response of 100 dipolar materials, solids as well as liquids, in the form of a plot of the exponents m versus $1 - n$ in the empirical expression (5). Each material is represented by a circle with the number referring to the index to be found in the original reference (Hill 1980). In addition, some acoustic absorption data are represented by triangles and mechanical modulus data by squares. The Debye response corresponds to the point 'D' in the top right-hand corner, the broken diagonal relates to symmetric peaks represented by the empirical expression of Cole and Cole, and that of Fuoss and Kirkwood. The top side corresponds to the empirical expression of Cole and Davidson and also to the Williams and Watts time-domain formula.

2.2. *Charge carrier systems*

Although the dielectric response is commonly associated with the orientation of permanent dipoles, it is undeniable that *hopping charges* of either electronic or ionic nature may give rise to very similar dielectric behaviour. The important distinction lies in the degree of localisation of these carriers: an electron or an ion confined to hopping between two preferred positions is indistinguishable from a dipole, as was seen on the example of Er-doped CaF_2, while an essentially different situation arises where the carrier is free to execute consecutive hops over finite paths, some of which may eventually extend all the way from one electrode to the other. Typical examples of this are electronic glassy and amorphous semiconductors and also most crystalline ionic materials, including the so-called fast ion conductors. In the case of hopping electronic systems it has been recognised for some time that the experimentally observed frequency dependence of the electrical conductivity followed equation (4) with

$$\sigma'(\omega) \propto \omega^n \qquad \text{with} \qquad n < 1, \text{typically} \simeq 0.8 \tag{11}$$

which corresponds exactly to the high-frequency limit of equation (10) in the dipolar case. Figure 6 shows the compilation of experimental data for a wide range of materials which include hopping electronic and ionic conductors; although some of the examples are evidently not in this category, the authors of the relevant papers felt that they were and the interpretation given is in terms of hopping conduction. Obvious examples of this are anthracene and β carotene.

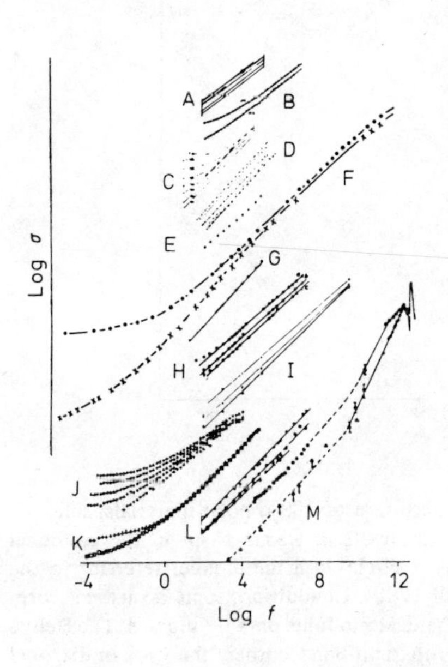

Figure 6. The compilation of AC conductivity data for a range of materials thought by the authors of the original papers to conduct by hopping charge movement. The plot is on a common frequency axis but data sets corresponding to different letters are displaced vertically for clarity. The symbols denote the following materials: (A) Single-crystal silicon in the impurity hopping range for electrons, 3–12 K; (B) single-crystal β alumina at 77 and 87 K – a Na^+ ion conductor; (C) glow-discharge-deposited amorphous silicon 84–295 K; (D) a range of chalcogenide glasses at 293 K; (E) single-crystal anthracene at 294 K with saline solution contacts; (F) single-crystal anthracene (xxx) and evaporated β carotene (···); (G) TNF–PVK; (H) and (I) oxide glasses; (J) evaporated silicon monoxide 211–297 K; (K) 9-layer stearic acid film with Al and Au electrodes in the dark (□□□) and with UV light (+++); (L) three amorphous chalcogenide samples; (M) two samples of As_2Se_3 showing quantum effects at very high frequencies. Source references and further details in Jonscher (1980a).

A strong conviction appears to have arisen among workers in the field of AC conductivity that the power law relationship of the type (11) constitutes a *proof* of hopping *electronic* conduction – not ionic – and many results which do not apply to electronic materials are incorrectly classified because of this. The reason for this is easy to find in the great popularity of amorphous semiconductor studies during the 1960s and the identification of this type of response with electronic hopping processes. What seems to have eluded everybody is the fact that there are many more non-electronic systems obeying the same relationship. The same misconception extends also to the field of interpretation: theories of AC conduction in hopping systems follow their own path which is equivalent to the accepted theories of dipolar systems and they appear not to recognise the need to take many-body processes into account in a comprehensive way that would encompass all dielectric materials.

A remarkable feature of the results shown in figure 6 is their relatively very narrow range of absolute values of conductivity, as may be seen in figure 7, where all the data are plotted on a common axis. It would be extremely difficult to explain why Na – β alumina and anthracene should fall in the same range if the number of available carriers were the dominant factor.

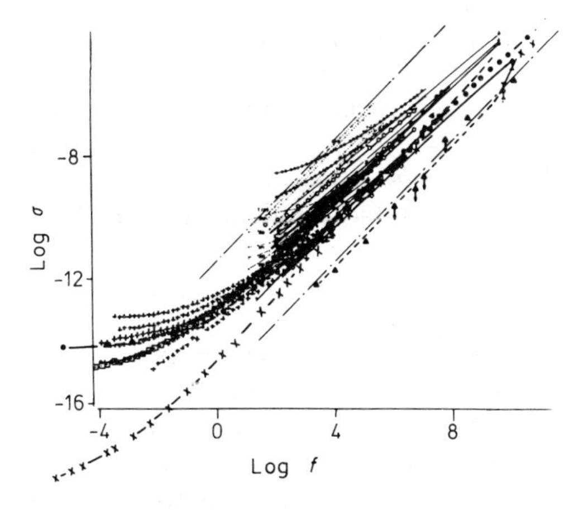

Figure 7. The same data as shown in figure 6 plotted in their correct relative positions on a common log $\sigma(\omega)$ axis (σ in $(\Omega\ \mathrm{cm})^{-1}$), showing the relatively very narrow range of AC conductivity for widely different materials. The sloping chain lines correspond to frequency-independent values of loss, $\chi'' = 10^{-3}$ and 10. From Jonscher (1980a).

Leaving aside the close similarity of the high-frequency responses for hopping charges and for dipoles, the other remarkable feature of all the data shown in figure 6 – with the exception of anthracene and β carotene – is the absence of any evidence of loss peaks down to the lowest frequencies which, in the conductivity presentation would correspond to slopes steeper than unity at the low-frequency side of the peak. Instead, all the data show either an unchanged slope down to the lowest available frequencies, or the slope decreases towards low frequencies as would be expected in the presence of a DC conductivity mechanism. Furthermore a close inspection of the data reveals that in some cases there is no visible saturation towards a value σ_0, but instead a definite small gradient persists, indicating a value of the exponent n close to but significantly different from zero. This will be taken up in the next section.

2.3. *Extreme values of the exponent n*

There is ample experimental evidence that the exponent n in the universal relation (10) or (11) can have values very close to unity, making the dielectric loss practically independent of frequency, sometimes over six or more decades. The frequency-independent loss is also virtually temperature-independent and it represents a very characteristic feature of all low-loss dielectrics. It appears that no solid dielectric has a loss descending below the limit of sensitivity of the best equipment, implying that any loss peaks that may be present are superimposed on a flat loss for which $n \to 1$ (A C Lynch and W Reddish, private communication). It is also our understanding that this is not the case in liquids, where the loss between loss peaks may descend below the detection limit. The observation of $n \to 1$ is facilitated by a lowering of the temperature, when other loss mechanisms tend to move to lower frequencies, and also by the purification of the samples which tends to remove additional 'extrinsic' dipolar and charge species. This process is therefore associated by us with the 'intrinsic' response of a 'perfect' dielectric lattice.

The other extreme value $n \to 0$ was mentioned in the previous section in the context of 'carrier-dominated' dielectrics. The implication here is that both the real and the imaginary components of χ follow the strongly dispersive relation (10) with a small value of n, maintaining the ratio $\chi''(\omega)/\chi'(\omega) = \cot(n\pi/2)$ constant and large, implying high losses (Jonscher 1978b). Many examples of this type of behaviour have been found and this type of response is more common, especially at high temperatures, than the classical DC limit in which $\chi''(\omega) \propto \omega^{-1}$ and $\chi'(\omega) \to \chi(0)$ = constant. Setting in equation (5) $s = 1 - p$, we may write the time-domain response corresponding to this regime:

$$i(t) \propto t^{-1+p}, \qquad p \lesssim 1, \qquad \omega \ll \omega_c \tag{12}$$

where ω_c now indicates the critical frequency at which the transition takes place from the high-frequency regime (10), corresponding to values of n nearer unity, to the low-frequency regime in which $\chi''(\omega) \propto \chi'(\omega) \propto \omega^{-p}$. We may say formally that in this dispersive regime $-p$ replaces m in the dipolar case, equation (8).

We stress that neither of the two cases $n \to 0$ and $n \to 1$ had been recognised previously as specific types of dielectric response and must have been consistently ignored when actually observed, since there was no corresponding theoretical interpretation and it is very difficult to see how these cases can be understood in terms of accepted models, unless one regards the low-frequency dispersion as a manifestation of the Maxwell–Wagner process. By contrast, in our universal approach both cases arise as natural limits of a comprehensive range of responses.

3. Classification of response types

Our wide-ranging study of the dielectric behaviour of various materials has led us to the formulation of a general classification of all known types of dielectric response in the wide frequency range below the inertial and quantum regimes. This is shown schematically in figure 8 which gives, in the upper row, the log χ' and log χ'' against log ω plots and in the lower row the corresponding complex susceptibility plots.

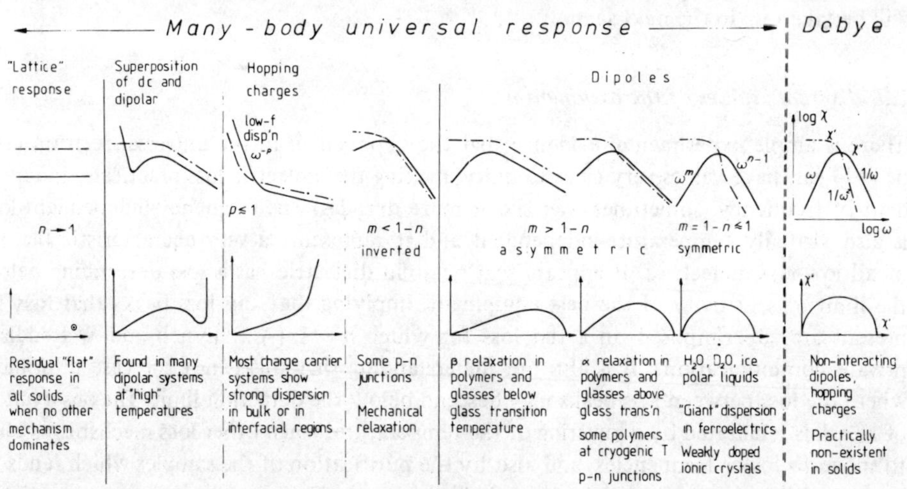

Figure 8. The general classification of all types of dielectric responses found in solids. The upper row gives the diagrammatic representation of the log $\chi'(\omega)$ (chain curve) and log $\chi''(\omega)$ (solid curve) against log ω, the lower row gives the corresponding complex susceptibility plots. Typical materials giving the various types of response are indicated. The extreme right gives the practically non-existent case of the Debye response; moving to the left we find increasingly broader loss peaks for dipolar systems, further to the left the charge carrier responses corresponding to the strong low-frequency dispersion and to DC conductivity. On the extreme left is the limiting case of 'flat' frequency- and temperature-independent loss.

On the right we see the ideal Debye response which is hardly ever seen in solids, since it corresponds to non-interacting charges or dipoles, while the least departure from the ideal Debye response necessitates a change of the slope of the $\log \chi' - \log \omega$ plot, as explained in the context of equation (10). Moving to the left, we go through a range of increasingly 'non-Debye' types of responses with increasing widths of the loss peaks, corresponding to progressively decreasing values of m and $1 - n$. Here belong the well known α and β peaks of glassy materials above and below the glass transition temperature. The 'near-Debye' peaks found in many ferroelectric systems at gigahertz frequencies where the 'giant' dispersion takes place as the systems cease to follow the high frequency, also in highly pure p–n junctions and in many liquids, are well approximated by the empirical Cole–Cole relation; symmetric peaks may also be represented by the Fuoss–Kirkwood formula. For asymmetric peaks the widely accepted formalisms are the Cole–Davidson and Williams–Watts expressions, the latter giving in the time domain

$$i(t) \propto t^{-n} \exp(-t/\tau)^{1-n} \tag{13}$$

where τ is a suitable relaxation time. It should be stressed, however, that the last two expressions remain one-parameter relations which give the slope corresponding to our exponent $m = 1$ only at low frequencies, corresponding to the upper side of the square in figure 5. This means that these empirical expressions lack the essential flexibility to reproduce the experimentally observed variety of shape functions of the dielectric responses. On a purely formal basis, without any claim to represent a physical model, the Havriliak–Negami formula does this, by combining the Cole–Cole and Cole–Davidson expressions.

We note the fact that a few dielectric systems show the 'inverted' form of loss peak in which the exponents are $m < 1 - n$, corresponding to the space below the diagonal in figure 5. By contrast, many mechanical loss data fall in this range (R M Hill, private communication).

Passing now to systems in which the polarisation is dominated by hopping electronic or ionic charge carriers, we note the absence of loss peak and its replacement by a region of strong low-frequency dispersion, obeying the Kramers–Kronig relations and contrasting strongly with the onset of DC conductions on top of some dipolar mechanism. We note in passing that one class of systems in which we consistently find DC rather than strongly dispersive behaviour is semiconductor p–n junctions and, significantly, these are systems in which there is no contribution from hopping charge carriers.

The extreme left of the diagram represents the limiting case of flat frequency- and temperature-independent loss, $n \to 1$.

4. Dielectric response at very low temperatures

One of the most significant aspects of the dielectric response of solids is that it cannot be 'frozen out' by a lowering of temperature to the millikelvin range – significant levels of loss and dispersion of $\chi'(\omega)$ persist despite the fact that all other forms of transport, except electronic tunnelling in some systems, cease to operate.

The dielectric response at very low temperature takes on one of two distinct forms. In some polymeric materials, especially those containing hydrogen bonds, e.g. slightly oxidised polyethylene, a distinctive loss peak persists down to millikelvin temperatures, its width generally not exceeding 2–3 times the Debye width, but the temperature dependence of ω_p is not activated and it may become temperature independent (Hill and Jonscher 1980). At the other extreme are the substantially flat responses found in many materials, and this may be taken as a tendency for $n \to 1$, although figure 2 shows an example of suprasil glass where clearly $m \to 0$. The flat frequency- and temperature-independent loss is found in such unmistakably ionic conductors as K^+–hollandite structures and Na^+–β alumina in which there are no sufficiently light tunnelling species, such as electrons or protons, so that the observed high dielectric activity must be due to the presence of some other non-activated mechanism.

5. The behaviour of p–n junctions in crystalline semiconductors

The p–n junctions in device-quality semiconductors represent some of the most structurally perfect and purest dielectric systems whose physical and chemical characterisation is much better than that of any other materials. In addition, p–n junctions have the unique advantage of not involving metallic interfaces to the dielectrically active region, since contact is made through monolithic crystalline semiconductor material of the extrinsic p and n regions on either side of the space charge region. This completely eliminates any uncertainties that may arise regarding the role of interfacial processes. The width of the space charge region may be altered by the application of a steady bias superimposed on the small signal and this provides an additional parameter in the study.

It is interesting to note that, despite their considerable technological significance, very few studies have been made of the frequency dependence of dielectric loss and all accepted interpretations involve the familiar concepts of Debye-like responses arising from charge generation and recombination in the space charge region. The Chelsea Dielectrics Group have pioneered detailed studies on a wide range of p–n junctions (Barsony and Jonscher 1978, Jonscher *et al* 1981) both as regards the range of frequencies, temperatures and junction types. We report responses approximating to Debye behaviour in silicon diodes with very low densities of localised levels in the space charge region, through broadened responses in high-energy electron bombarded junctions and on to 'inverted' very asymmetric peaks in silicon and laser GaAs diodes.

One significant feature of junction response is that the application of a reverse bias appears to reduce the level of loss, while retaining its spectral shape – this is a rather unexpected result. Furthermore, the fact that both silicon and GaAs have very similar loss characteristics, both as regards their spectral shape and the absolute values of ω_p, again calls for detailed consideration in the light of a new approach, since the accepted interpretations appear not to be consistent with these observations.

We have already noted the complete absence of strong low-frequency dispersion from p–n junction characteristics, which we believe to be due to the fact that no hopping charge carriers are involved in these systems. We believe that a proper understanding of the dielectric response of p–n junctions must take into account the non-exponential nature of the generation/recombination processes – a hitherto unsuspected phenomenon.

6. The universality of dielectric response

The preceding discussion shows clearly that the dielectric behaviour of virtually all solids follows the 'universal' law which may be expressed most conveniently by the power law time-domain response given by equation (5), with the exponent s taking on well defined values in specified ranges of time, between 'short times' of the order of 10^{-11} s and the longest experimentally accessible times. A typical 'elementary' or 'simple' relaxation behaviour found in very many materials is shown schematically in figure 1, where we differentiate between dipolar and charge-dominated systems.

We have already mentioned the fact that it would be unrealistic to expect all materials to show clearly 'simple' behaviours – superpositions of two or more mechanisms partially overlapping in frequency, each of the 'simple' type, are frequently seen. Indeed, what is surprising is that so many examples exist where a 'simple' law is seen over the entire available frequency range. This we take as an indication that the proposed 'universal' behaviour is a strong characteristic of most dielectric systems and this justifies our attempt to provide a unifying theory covering all these situations.

We propose, therefore, that the universal response as defined by equation (5) and the corresponding frequency-domain response, is found in all solids, regardless of:

(a) their physical structure and long-range order – single crystal, polycrystalline, amorphous and glassy;
(b) the prevailing type of bonding – covalent, ionic, molecular;
(c) the nature of the polarising species – dipoles, hopping electrons, polarons and ions;
(d) the geometrical configuration – from 'bulk' samples to monomolecular layers and very thin barrier regions, for planar and intricate geometries, continuous and granular media.

The remarkably general applicability of the universal response has led us to the conclusion that we should look for a correspondingly general model of dielectric polarisation that would be applicable to all these very different conditions and would be capable of accounting for the essential similarity of power law relations in all these cases. In doing this we are conscious of the many theories of dielectric response which have been proposed in different branches of the subject, to account for the observed behaviour of specific types of materials. We do not propose to review them in the present paper since they are treated in some detail elsewhere (Jonscher 1977b, c, 1980a) and it will be sufficient merely to state here the principal conclusions of this survey. This is that, even though some of these theories may be plausibly applicable within the context of the various materials for which they were specifically developed, the overriding fact remains that the existence of a universality of experimental behaviour demands a higher-order explanation why all the different theories should ultimately give the same end result in the form of a power law.

A serious difficulty with many of the currently accepted theories is their arbitrariness in the choice of the disposable parameters in the analysis. The classical example here is the distribution of relaxation times where it is impossible to prove that the distribution function required to explain the results actually exists or is even remotely plausible on physical grounds. The possibility exists that some of the material responses may be theoretically explicable on the basis of one of the existing theories, while others may require a more general interpretation. For example, the near-Debye results may require only a small modification of the classical model. Without wishing to resolve this point, it is suggested here that the majority of dielectrics depart sufficiently drastically from the Debye response to require a completely new approach to the theory.

We propose that a theory based on many-body interactions is well capable of accounting for the totality of the observed results. Such interactions affect all particles in condensed matter on account of their close proximity, but the energies involved are very low in comparison with, for example, typical one-electron excitations considered in the energy band model of solids. Many-body interactions are, therefore, of secondary importance in steady-state hopping transport, in which the rate-limiting transitions involve large energies, or in free-electron transport which is dominated by thermal velocities. By comparison, dielectric processes involve small, low-energy displacements of large numbers of particles and many-body effects become dominant.

The failure of most hitherto accepted theories to provide a satisfactory overall account of the dielectric behaviour of all solids lies in their attempts to substitute distributions of one-particle parameters for many-body interactions. In the case of some correlation function approaches the importance of many-body interactions is recognised, but the analytical methods employed would require a large number of terms to be summed to give a full description and this is proving to be very difficult. On the other hand, the alternative of using only a few terms gives very poor convergence and the results are impossible to interpret.

7. The criteria of universal behaviour

Before formulating the principle of the many-body approach we have to establish the common features which apply to all the diverse materials showing the universal response. We proposed some time ago that the only such features were the suddenness of the transitions being made by the individual charges or dipoles, and the presence of interactions between these charges or dipoles (Jonscher 1977 a, b, c). The abruptness of transitions is a characteristic feature of all solids where charges occupy preferred sites and dipoles preferred orientations and these are separated by potential barriers which prevent smooth transitions being made of the type originally envisaged by Debye for his dipoles 'floating' in a viscous medium.

In the present approach we emphasise the difference in time scale between the abruptness of the individual transitions and the slow nature of the cooperative adjustment of the surrounding medium following each such transition. The new approach is directly linked to this combination of two entirely different time scales for individual and collective excitations and it represents a clear break with earlier treatments.

One other essentially common feature of all the material systems under investigation is the presence of *disorder* as an intrinsic characteristic of all polarisation mechanisms, whether they are due to orientation of permanent dipoles or to hopping charge carriers (Jonscher *et al* 1980). This disorder is present at several levels, beginning with the most obvious arising from the fact that permanent dipole vectors in any orientationally polarisable medium are necessarily randomly oriented in space, even if their positions in the lattice should be perfectly regular. The second level of disorder arises from the random distribution of the dipoles or charges themselves, which means that the dipole–dipole interactions are inevitably different between different pairs, and the same is true of the carrier–carrier interactions. Finally we have the third level of disorder in the atomic, ionic or molecular lattice in which the permanent dipoles or extrinsic charges are distributed, since the properties of this lattice are influenced locally by the presence of the dipoles or charges.

All this implies that dielectric systems represent a considerable degree of disorder and this would be expected to cause resemblance to the well known amorphous materials which exhibit many features that are unfamiliar in the context of perfectly ordered solids (Mott and Davis 1979). Among these features are the anomaly of specific heats and the absence of clearly defined activation energies for hopping transport. We conclude therefore that, by virtue of their inherent disorder, dielectric materials show a certain 'softness', in contrast with the 'hardness' of perfect order which requires a large amount of energy to be broken.

8. The infrared divergence model

It has been pointed out by K L Ngai (Ngai *et al* 1979) that the criteria for the onset of the universal dielectric response mentioned above, are equivalent to those which determine the infrared divergence (IRD) or the x-ray anomaly, whose time evolution follows the same power law t^{-n}. The requirements for the occurrence of these processes in an interactive system are that:

(a) the system should be capable of being excited by a sudden switching of potential, and

(b) there should exist lower-energy excitations forming a continuous spectrum of constant density in energy which become excited by this potential and which cause the emission of excitations with ever decreasing energy and extending in time to infinity.

The presence of disorder leads to interactions between constitutent elements in the system, in that the energy of each unit is influenced by the positions or orientations of all other units. This means that the system is essentially nonlinear and the solutions of the resulting nonlinear Schrödinger equation lead to the appearance of IRD. Self-consistent solutions possessing the required properties for the appearance of IRD were obtained by Ngai who was thus able to show that the ubiquitous t^{-n} law generally observed in dielectric relaxation was consistent with this model (Ngai 1980, Ngai and White 1979). The limitation of the Ngai theory lies in the fact that it represents a one-parameter treatment which is equivalent to the Williams–Watts equation (13). The consequence of this is the inability to account for values of our exponent m which are different from unity.

The precise form of the distribution of the density of these states in energy is not known exactly at present, but it is reasonable to expect (Jonscher *et al* 1980) that it has a form shown schematically in figure 9. Although this is far from constant in energy, as

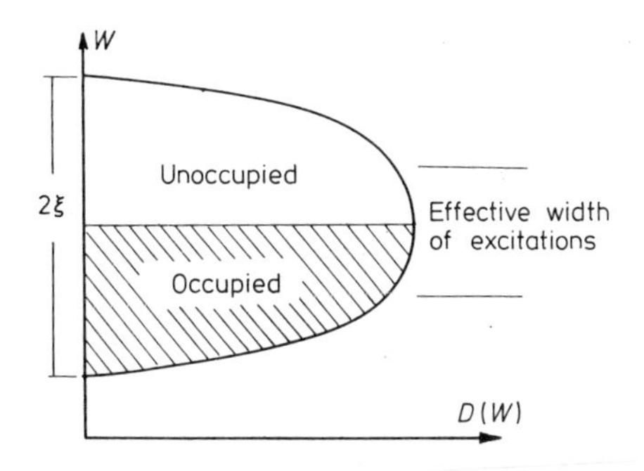

Figure 9. The distribution function in energy of the density of correlated states, showing the 'occupancy' of these states in equilibrium and the effective width of excitations for which the density is approximately independent of energy (From Jonscher *et al* 1980).

would be required by (b) above, it may be assumed that the middle range over which most of the effective excitations take place is sufficiently close to a constant distribution to satisfy the requirements of the theory.

The band of correlated states arising from the interactions is believed to be narrow compared with typical thermal energy kT at normal temperatures and this means that these states are not in thermal equilibrium with the phonons, since two phonons at least would be required to couple to these very-low-energy excitations. This explains why the universal response is not 'washed out' at normal temperatures. If the width of the band of correlated states is 2ζ, corresponding to frequencies of the order $10^{12}\ \mathrm{s}^{-1}$, and the typical switching potential due to the abrupt dipolar or charge carrier transitions is V_f, then the exponent n in the universal relation is given by

$$n = |V_f/2\zeta|^2 \tag{14}$$

Since only values of $V_f < 2\zeta$ are physically significant in determining the dielectric response, this explains why the observed range of n lies between 0 and 1.

9. The Dissado – Hill theory

The IRD model was further developed independently in a series of papers by Dissado and Hill (1979, 1980, to be referred to as D&H) who have introduced important new concepts to bring the theory into better agreement with the empirically determined behaviour of dielectrics, especially the presence of two independent mechanisms giving rise to the exponents m and n, the unfreezability of dielectric response and the temperature dependence of ω_p. This work has also led to important advancements in our physical understanding of the various many-body interactions involved in dielectric relaxation.

The D&H theory is based on a two-level potential well model describing the energy of a large number of individual systems in an interactive dielectric medium, see figure 10. The splitting $2B_{\mathrm{eff}}$ of the bottoms of the potential wells is determined by the local potential configuration and also by the actual occupancy of the two minima, i.e. by the net dipole moment M, as a consequence of the interaction. There is also a contribution from any external electric field E, so that:

$$B_{\mathrm{eff}} = B + kT_c M + QE \tag{15}$$

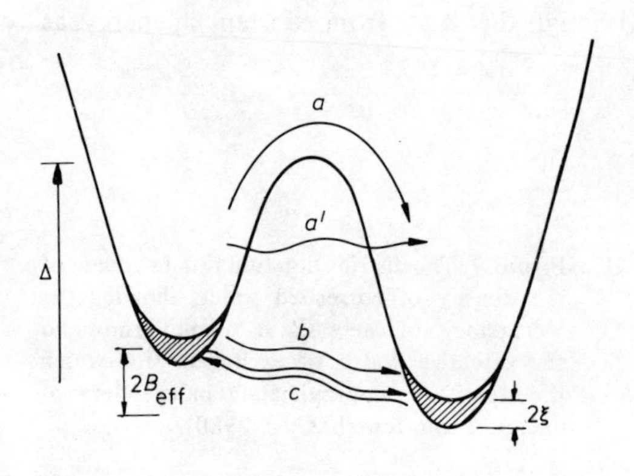

Figure 10. The double potential well for the two-level model of Dissado and Hill representing the energy of a large number of individual systems in an interactive dielectric medium. The two minima correspond to two orientations of the dipoles with respect to an external field, the splitting $2B_{\mathrm{eff}}$ being determined by their occupancy and by the field. The shaded regions near the bottoms of the wells denote the correlated states of width 2ζ in energy. Δ denotes the potential barrier separating the minima for thermal excitation. The significance of the arrows is indicated in the text.

where T_c denotes the strength of the interaction and Q is the dipole moment corresponding to the transition. This gives a nonlinear equation for the equilibrium dipole moment

$$M_e = \tanh\left\{\frac{B + kT_cM_e + QE}{kT}\right\}. \tag{16}$$

The return of the system to equilibrium after a perturbation M' from M_e takes place through three types of transitions. Those marked by the arrow a in figure 10 take a dipole over the barrier Δ from one orientation to the other with a corresponding input of thermal energy. The transition may be partly tunnel-assisted, or may be entirely tunnelling, as indicated by the arrow a' but this is only possible for light particles, such as electrons or protons, excluding heavier particles. These transitions are denoted as 'large', since a single particle makes a large spatial jump and D&H have characterised this behaviour in detail for the special case of order–disorder transitions. The rate equation for the 'large' transitions is then found to be

$$\frac{dM'}{dt} = -\omega_p M' \tag{17}$$

where the frequency ω_p is given by

$$\omega_p = \nu_A \;\{1 - (T_c/T)(1 - M_e^2)\}\; \cosh\left\{\frac{B + kT_cM_e}{kT}\right\} \tag{18}$$

where ν_A is a thermally activated jump frequency. This corresponds to the classical Debye relaxation of the dipolar system according to equation (3), but in the present case it is valid only at times $t \simeq 1/\omega_p$, i.e. in the neighbourhood of the reciprocal loss peak frequency.

Two other types of tunnelling transitions are envisaged in the D&H theory and will be referred to as 'small' transitions. Tunnelling in these terms refers to *cooperative local* readjustment of atomic positions such that the entire local group of atoms is transferred from one *local* configuration to another. Activated tunnelling refers to such movements which are assisted by thermal means, and are therefore irreversible. Configurational tunnelling refers to the same motions which are not thermally assisted. In the absence of group–group interactions they are oscillatory and conserve local memory. In the presence of group–group interactions the local potential in which such motions occur is continuously altered by the presence of such motions elsewhere. This leads inevitably to the loss of memory of any local oscillations and consequently to a decay. The energy is thus redistributed over the assembly of groups. The intervention of thermal motions is not required for this process.

With reference to the potential well of figure 10 we note the fact that this well is described by a potential which is quartic, i.e. 4th power, in the displacements. In the presence of interactions between systems these terms lead directly to a non-exponential time response under transient excitation (Halperin and Hohenberg 1976, Halperin *et al* 1972). One type of these transitions, denoted by the arrow b in figure 10 corresponds to configurational tunnelling in which small adjustments of positions of large numbers of dipoles or charges give a net change of dipole moment equivalent to that of a large transition. These 'flip' transitions do not involve any thermal excitation and their time evolution is obtained from second-order perturbation theory at times long compared to

$1/\zeta$ but short compared to $1/\omega_p$

$$i(t) \propto (\zeta t)^{-n}, \qquad 1/\zeta \ll t \ll 1/\omega_p \tag{19}$$

This is the equivalent of the expression obtained by Ngai.

The 'large' and 'small' transitions are contrasted in figure 11 which shows an idealised one-dimensional 'lattice' of oppositely charged ions. In the upper diagram the ions occupy regular positions in the lattice, but one site is unoccupied, leaving a net dipole moment indicated by the arrow. The vacant site may be filled by a 'large' transition of the nearest like ion and this corresponds to a change of the dipole moment to the value indicated by the lower arrow. The second diagram considers a hypothetical 'disordered' lattice in which the positive and negative ions are slightly displaced from the regular positions as indicated. If now the individual ions should take up equivalent opposite positions through very small displacements, the combined effect for all particles is a change of the dipole moment by a comparable amount to that involved in the 'large' transition.

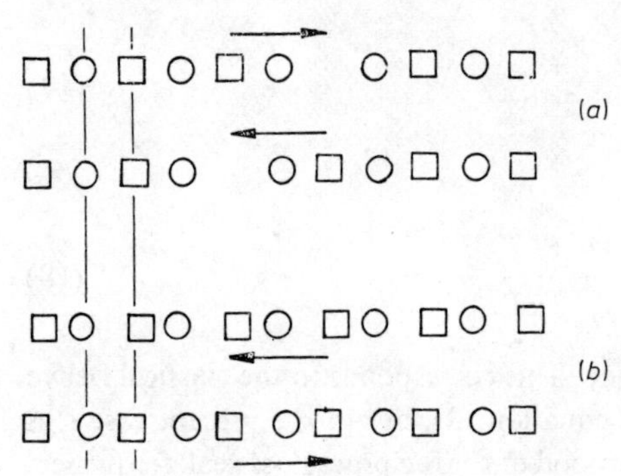

Figure 11. A schematic representation of (*a*) a 'large' transition, and (*b*) 'small' transitions, in a one-dimensional 'lattice', as described in the text. The arrows indicate the net dipole moments in each case and the displacements in the lower diagram add up to the same value as the large transition in the upper diagram.

The other type of transition envisaged in the D&H theory is represented by the arrows *c* and corresponds to simultaneous flip transitions in opposite senses at different points in the system, which therefore leave the net dipole moment unchanged. These 'flip-flop' transitions give rise to a redistribution of excitation energy among the correlated states and they therefore influence the rates of the other transitions, leading ultimately to the time law

$$i(t) \propto t^{-1-m}, \qquad t \ll 1/\omega_p. \tag{20}$$

The flip-flop transitions correspond to fluctuations on the system and they are directly related to noise in the dielectric system, especially to the ubiquitous '1/f noise', thereby providing a quantitative connection between this noise and the measurable dielectric parameters (Hill 1981). The exponent m in this case is related to the width 2η of the flip-flop excitations and to the average potential V_{ff} of a flip-flop process, by the relation

$$m = |V_{ff}/2\eta|^2 \tag{21}$$

which is analogous to equation (14) for n.

The D&H theory provides a complete description in closed form of the time and frequency response of dielectrics in terms of a *shape function* $F(\omega/\omega_p)$ and an *amplitude function* of the susceptibility which is given by:

$$|\chi| \propto \frac{1-M_e}{kT}\omega_p \; \{1-(T_c/T)(1-M_e^2)\}^{-1}. \tag{22}$$

The shape function of the reduced frequency variable is determined by the two independent parameters m and n which are sufficient to account for all observed forms of response. The fact that the complete dielectric behaviour may be represented by the product

$$\tilde{\chi}(\omega) = |\chi| F(\omega/\omega_p) \tag{23}$$

constitutes the theoretical justification for the often used technique of normalisation of dielectric data corresponding to different temperatures, which is also known as the principle of time–temperature superposition.

10. Discussion of the D&H theory

This theory, developing the IRD model initiated by Ngai, provides a very powerful means of understanding the nature of relaxation processes in dielectrics which are governed by the interactive processes and this means in practically all solids. Several aspects of this theory are amenable to direct experimental tests, while other features are currently being developed. We shall mention below some of the principal points.

10.1. Relationship between $|\chi|$ *and* ω_p

A completely new relation may be derived directly from the D&H theory, using equations (18) and (22) (Hill and Dissado 1979). In materials in which T_c is nearly independent of temperature, indicating the presence of a phase transition, the very simple relationship is obtained:

$$\chi''_m \propto \omega_p^n \qquad T \gtrsim T_c \tag{24}$$

$$\chi''_m \propto \omega_p^{-1+m} \qquad T \simeq T_c \tag{25}$$

where χ''_m denotes the maximum amplitude of the loss peak. These previously unsuspected relations have been clearly confirmed in a number of cases and they constitute a startling validation of the proposed approach.

10.2. Limiting cases of high ω_p

Where the loss peak frequency corresponding to the large transitions in figure 10 is sufficiently high, in the range $10^8 - 10^9$ Hz, the flip-flop transitions do not have time to intervene, since they require times of the order $1/\eta \simeq 10^{-9}$ s to develop. In this case the limiting low-frequency response is proportional to ω, i.e. $m = 1$, and this is confirmed by the several examples in figure 5 which fall on the top side of the square, the majority of which correspond to loss peaks in the very high frequency range.

10.3. Significance of n *and* m

The exponents determined by equations (14) and (21) are also directly related to the correlation of the flip and flip-flop transitions, respectively, the value unity corresponding in each case to full correlation, the value zero to complete lack of correlation. In this context we recognise the 'flat' loss as the limit of strongly correlated flip transitions, each flip being necessarily followed by another and this is expected to be the case in 'rigid' systems which do not admit of adjustments between neighbouring dipoles.

10.4. Temperature dependence of the shape function

The shape of the frequency dependence function, $F(\omega/\omega_p)$, especially its loss component, is in general remarkably independent of temperature. However, in the region of a structural transition the peak becomes narrower, i.e. more Debye-like as the temperature increases. The sharpening of the peak corresponds to an increase of m and a decrease of n with rising temperature. The latter may be understood in terms of the increasing structural disorder reducing the correlation of the flip transitions, while the same process may be expected to increase the correlation of the flip-flop transitions which are enhanced by an increase of disorder. The limiting form of this type of behaviour is seen in strongly associated liquids such as water, which approximate to the Debye response.

10.5. 'Flat' loss in solids

The fact that solids tend to show an underlying trend of frequency-independent loss, while liquids do not, may be seen as the consequence of an inevitable high level of disorder in liquids which makes strong correlations of flip transitions very unlikely.

10.6. Response at very low temperatures

The observed unfreezability of the dielectric response is very difficult to explain in the context of the conventional dielectric theories, at least where the mobile charge species are heavier than electrons or protons, for which one-particle quantum mechanical tunnelling may plausibly be invoked. The present many-body theory of configurational tunnelling provides a simple interpretation of the observed dual nature of the response – either with or without a loss peak. Where a peak is observed one has to invoke the presence of 'large' single-particle tunnelling, such as might be expected in hydrogen-bonded systems, e.g. carboxylic groups in slightly oxidised polymers. Alternatively, thermally activated tunnelling may lead to a loss peak.

Where a loss peak arising from thermally assisted large transitions is present at high temperatures, this peak must disappear below a certain temperature at which the thermal energy becomes insufficient to cause the excitation.

10.7. Semiconductor p–n junctions

The dielectric response of p–n junctions, otherwise so similar to that of many other dielectric systems with dipolar orientations, poses some very interesting problems of interpretation, since we are not able to use the normal concepts of dipoles in the case of deep levels in the junction space charge region (Jonscher *et al* 1981a). At the same time, the usual interpretation in terms of distributions of generation–recombination times at deep levels in the forbidden gap does not seem to explain adequately the observed 'universal' nature of the power law responses. We propose to consider the effect of many-body interactions on these generation – recombination processes, which should change their temporal evolution from the normally assumed exponential dependence to power law type.

11. Strong low-frequency dispersion

At the present time we have no rigorous theory of this phenomenon which is observed generally in carrier-dominated systems with hopping electronic or ionic charges, as distinct from dipolar systems. Formally, this behaviour may be associated with the change of the exponent m in the D&H theory to $-p$ in the long-time response. This change is equivalent to the condition that the flip-flop transitions should *enhance* the flip transitions, rather than compete with them, as in the dipolar case. It is possible to envisage how this arises physically in a system of charges moving on a network of 'easy' hopping transitions interrupted occasionally by more difficult hops. The traversal of individual paths by charges is equivalent to flip transitions involving 'giant dipoles' (Jonscher 1978b) and the flip-flop transitions are then represented by hopping adjustments between neighbouring paths which simulate the bypassing of difficult barriers. In this manner we see the difference between dipolar loss peaks and the strong dispersion as arising from the three-dimensional nature of the hopping transport which changes the role of flip-flop transitions in comparison with the dipolar situation.

It is relevant to point out here that the activation energy for the low-frequency dispersion is very similar to that of the DC conductivity, confirming the similarity of the charge carriers involved in both processes.

12. Conclusions

This review has shown the existence of a definite universality of the dielectric behaviour of a very wide range of solids, regardless of their physical and chemical properties. The resulting general classification of various types of dielectric responses shows at least two limits not previously recognised as specific forms, *viz* the flat or frequency- and temperature-independent loss which is common to all solids as the limit when other dipolar or charge species have been eliminated, and the strong low-frequency dispersion in carrier-dominated systems. These two types correspond to natural limiting cases of our universal form of dielectric response and they are very difficult to justify in the accepted interpretations of dielectric behaviour.

We have concluded that the criteria for the occurrence of the universal behaviour are closely similar to those which would lead to the appearance of infrared divergence, with exactly the same type of behaviour as the 'short-time' dielectric response. This has led Ngai and also Dissado and Hill to develop models of IRD-like processes in application to the universal dielectric response, and the latter have succeeded, in particular, in obtaining a very satisfactory agreement with the general shape of the frequency dependence of dipolar systems. The essential feature of these models is the cooperative nature of dipolar and charge carrier movements, giving rise to the appearance of a distribution of correlated states forming a narrow band in energy. These states are not part of the conventional one-particle band theory of solids and they arise from the presence of disorder and of interactions between charges.

In the present interpretation, the loss peak in the frequency domain is considered to consist of three regions, with low- and high-frequency branches having power law dependence arising from the dominance of cooperative many-body processes, while the middle range between them reflects the predominance of classical Debye-like response arising from one-particle transitions over a potential barrier. This interpretation allows us, therefore, to retain the physical significance of the loss peak in terms of Debye-like processes, while at the same time explaining the often considerable departures from the ideal Debye shape in terms of many-body interactions.

The dielectric response is an extremely sensitive probe for interactive processes, since it involves excitations at very much lower energies than any encountered in steady state transport. This means that our knowledge of the detailed nature of these states remains as yet relatively rudimentary but we are gaining a much better insight into this subject by combining our theoretical approach with the unrivalled wealth of available experimental data on dielectrics.

The new approach has already proved to be a powerful tool for the understanding of the manyfold aspects of the dielectric response of solids, interpreting existing data and predicting new relationships which are then capable of being verified. By transferring the emphasis from one-particle processes to cooperative many-body interactions involving defects and disorder, the new approach is capable of explaining the otherwise very surprising experimental result implied in figure 6, i.e. the relative narrowness of the absolute range of values of dielectric loss, compared with the very much wider range of DC conductivity. It may reasonably be said that the lowest level of loss tangent in dielectrics, $\tan \delta = \epsilon''/\epsilon'$ is of the order of 10^{-5}, typically limited by the 'flat' loss, while the highest normally encountered value is of the order 1, implying a ratio of 10^5. We suggest that the explanation of this small range found in very different materials lies in the fact that the lower limit of loss is dictated by the local interactions at defects and the density of these is much less system-dependent than the density or mobility of charge carriers which are responsible for the DC conductivity, which can vary therefore by very many orders of magnitude.

The new approach provides a much more coherent and physically plausible theoretical picture of the nature of dielectric polarisation processes in solids than the diverse interpretations that have been accepted hitherto, none of which is capable of covering self-consistently the entire range of materials. It is clear, moreover, that progress currently being made in the understanding of the dielectric response is paving the way for rapid advances in many other fields of solid state physics, such as mechanical relaxation, noise theory, generation–recombination processes in semiconductors and chemical reaction kinetics, all of which entail many-body interactions. The advantage which dielectrics have over these other fields lies in the very large volume of experimental data, covering not only many materials but, in particular, an extremely wide range of frequencies which makes it difficult to 'fit' results with the customary exponential functions.

Acknowledgments

It is a pleasure to record my indebtedness to my colleagues at the Chelsea Dielectrics Group, in particular to Dr L A Dissado and Dr R M Hill whose support as theorists in the development of the concept and details of the universal response was invaluable. Science Research Council funding for various aspects of the work of the Group is also acknowledged.

References

Barsony I and Jonscher A K 1978 *Solid State Electron.* **21** 471
Dissado L A and Hill R M 1978 *Nature* **279** 685
— 1980 *Phil. Mag.* **B 41** 625

Fontanella J, Jones D L and Andeen C 1978 *Phys. Rev.* **B 18** 4454
Halperin B I and Hohenberg P C 1977 *Rev. Mod. Phys.* **49** 435
Halperin B I, Hohenberg P C and Shang-Keng Ma 1972 *Phys. Rev. Lett.* **29** 1548
Hill R M 1978 *Nature* **275** 96
— 1980 *J. Mater. Sci.* in the press
— 1981 to be published
Hill R M and Dissado L A 1979 *Nature* **281** 286
Hill R M and Jonscher A K 1980, this conference, unpublished
Jonscher A L 1975a *Colloid Polym. Sci.* 253, 231
— 1975b *Nature* **256** 5518
— 1977a *Nature* **267** 673
— 1977b *Phys. Status Solidi b* **83** 585
— 1977c *Phys. Status Solidi b* **84** 159
— 1978a *Thin Solid Films* **50** 187
— 1978b *Phil. Mag.* **B 38** 587
— 1980a *Physics of Thin Films* ed. M H Francombe and G Hass (London: Academic Press) vol 11, p205
— 1980b *Trans. IEEE E I,* to be published
— 1980c *J. Phys. D: Appl. Phys.* **13** L89
Jonscher A K, Dissado L A and Hill R M 1980 *Phys. Status Solidi b* **102** in the press
Jonscher A K, Favaron J, Charoensiriwatana V, Loh C-K and StCricq B 1981 this conference, to be published
Mott N F and Davis E A 1979 *Electronic Processes in Non-Crystalline materials* (Oxford: Oxford University Press)
Ngai K L 1979 *Comments on Solid State Physics* vol 9 no 4 pp127–140
— 1980 *Comments on Solid State Physics* vol 9 no 5 pp141–156
Ngai K L, Jonscher A K and White C T 1979 *Nature* **277** 185
Ngai K L and White C T 1980 *Phys. Rev.*

Inst Phys. Conf. Ser. No. 58
Invited paper presented at Physics of Dielectric Solids, 8–11 September 1980, Canterbury

Dielectric, mechanical and NMR relaxation

D W McCall
Bell Telephone Laboratories, Murray Hill, New Jersey, 07974 USA

1. Survey of relaxation studies

The commonality of molecular mechanisms underlying dielectric, mechanical†, NMR and other relaxation phenomena has been established for some time. Correlation frequency–temperature loci and relaxation strength analyses have provided a plausible basis for understanding specific group motions but explicit models for more general chain motions have been less satisfactory. In this paper we will survey representative results (see also McCall (1971) and McBrierty and Douglass (1980)). Although the data refer mainly to organic high polymers, similar motions and relaxations are observed in other materials.

Thus, for example, methyl group rotations can be said to be well characterised in general and there seems to be no ambiguity. The correlation frequency can be tracked over wide ranges in temperature by NMR and it is invisible in dielectric spectra as one would expect. Polypropylene, polyisobutylene (PIB), polymethylmethacrylate (PMMA), branched polyethylene, indeed, all methyl-containing polymers and other organic compounds, exhibit methyl relaxations. These relaxations occur at low temperatures and have small activation energies, 2–3 kcal mol^{-1}. The activation energies increase as steric hindrances are introduced into the local structure. Thus, PIB with two methyls on alternate chain carbons has an activation energy about twice that of polypropylene, which has only one methyl on alternate chain carbons. A consistent pattern emerges (McCall 1971).

Dynamic mechanical studies also exhibit relaxation behaviour corresponding to methyl rotation. This result implies that the potential wells that govern the methyl group motions do not show simply three-fold symmetry, and minima of different energies must be present (McCall 1966).

Other specific group motions can also be studied and assigned. The ester group motion has been studied in the methyl, ethyl, . . . series of the methacrylates and becomes virtually coincident with the glass transition when the ester group is butyl. Dielectric intensity yields a dipole moment of the correct magnitude. It is interesting to note that these ester relaxations are dominant in the dielectric spectra, exceeding the glass transition in strength. A curious and as yet unexplained feature is the NMR insensitivity to the ester motion in polyvinyl acetate (Hoch *et al* 1971).

Specific chain motions are also amenable to study and NMR has played a leading role in this area. The important first-order transition just below room temperature has been

†Dynamic mechanical relaxation studies are analogous to dielectric relaxation studies. A periodic mechanical stress is applied to a specimen and the in-phase and out-of-phase response is measured. Dynamic mechanical results often correlate with dielectric results but the theoretical interpretation of the mechanical effects is much less developed.

0305-2346/81/0058-0046 $01.50

shown to correspond to main chain rotation about the long chain (helix) axis in polytetrafluoroethylene (PTFE) (McCall *et al* 1967). Other, more subtle, conclusions can also be drawn concerning this relaxation and others in PTFE (McBrierty *et al* 1978). Linear polyethylene also exhibits a main chain rotation of this type (McCall and Douglass 1965).

As a general matter, however, linear polymer chain motions are complex and a complete understanding has yet to be developed. Low-temperature crystalline effects in polyethylene have been subjected to diverse analyses in terms of crystal defects (Hoffman *et al* 1966).

The glass transition is probably the most dramatic feature of relaxation spectra, and it is the least understood even though it is the most studied. When relaxation results are compared for glycerol, ortho-terphenyl, glassy $Ca(NO_3)_2\,4H_2O$, rubber and other polymers, the similarities are so striking that one concludes that local molecular structural details must be secondary to the overall behaviour. Certainly there is no unique characteristic ascribable to chain structures. The glass transition is probably best thought of as the natural transition from a rigid glass to a viscous melt, in some ways analogous to a melting transition. The molecular motion becomes long range above T_g but there are certainly motions of some sort that persist below the glass transition temperature. The latter point is revealed by examining the NMR $T_{1\rho}$ data below T_g. No break corresponding to T_g is observed (McCall 1971). The great breadth of the distribution of correlation frequencies is an essential feature of glass transition phenomena that must also be explained.

Glass transition phenomena in linear chain polymers are complicated and their interpretation is controversial. Relaxation data suggest that the amorphous regions of semicrystalline polymers undergo the glass transition in two stages. Disputes concerning which stage corresponds to the 'true' glass transition have been carried on enthusiastically for many years. Curiously, PTFE appears to be exceptional in the sense that it displays a 'normal' glass transition (McCall *et al* 1967).

In conclusion, it can be said that a great deal is known about molecular motions, of specific group relaxations, but knowledge becomes limited as the sizes of the relaxing units get larger. In a sense, the least important relaxations are well understood and the most important relaxations are least understood. We may console ourselves with the thought that melting points are difficult to describe theoretically as well.

2. Recent advances

The foregoing survey is based entirely on macroscopic relaxation results in which the experimental probes do not selectively excite specific parts of the systems. In NMR it is often possible to observe the crystalline and amorphous regions separately because the nuclei in the two regions relax at substantially different rates. Even so, there has been no high-resolution method for studying chemically distinct nuclei until recently. Spin diffusion, which occurs by mutual spin flips of neighbouring nuclei, tends to emphasise the fastest relaxation (McBrierty and Douglass 1980). Thus in linear paraffins, methyl rotations dominate proton relaxation for the entire specimen. That is, the methyl protons, which relax rapidly, provide a relaxation path for the methylene protons, which relax slowly. The effect is transmitted along substantial lengths of chain, e.g. $C_{94}H_{190}$.

In favourable circumstances it is possible to study polymer compatibility by following the effects of spin diffusion. The distance over which spin diffusion occurs depends upon the magnitude of the relaxation time under study. Mixtures (melt blended) of PMMA and a copolymer of styrene and acrylonitrile (PSAN) have been studied to see if the α methyl relaxation of PMMA influences the phenyl proton relaxation in the PSAN. The diffusion length for T_1 is about 150 Å and that for $T_{1\rho}$ is about 15 Å. Analysis shows that the level of heterogeneity is between 15 Å and 150 Å (McBrierty *et al* 1978).

As a result of recent advances in technique, it is now possible to obtain high-resolution NMR spectra of solids. Dipolar broadening is eliminated by means of sequences of exciting pulses and the shift anisotropy is eliminated by spinning the sample about an axis inclined to the static field by the 'magic' angle. It thus becomes possible to study relaxation of side group nuclei and main chain nuclei independently. Carbon-13 studies are particularly attractive. The interpretation of the $T_{1\rho}$ values obtained in these experiments is still under study (McBrierty and Douglass 1980).

The following examples illustrate some of the results recently obtained by high-resolution NMR in solid polymers. In poly(dimethyl phenylene oxide) the protonated ring carbons are not equivalent, implying conformational constraint preferences in the solid polymer that are not present in solution (Schaefer *et al* 1977). In PMMA, the tertiary chain carbon has a very short relaxation time T_1 (Shaefer *et al* 1977). This has been interpreted as evidence of a 'crankshaft' motion of the polymer chain. In the spectrum of an epoxy below and above T_g, only one line, corresponding to aromatic protons of diglycidyl ether of bis phenyl A (DGEBPA), changes as the sample is heated from well below to well above T_g (Garroway *et al* 1978). Again, conformational preferences are suggested.

This is a vast field that is only now beginning to open. The promise of high-resolution information in solids is most attractive if interpretations can be worked out that are satisfying and soundly based. This has not yet been achieved.

Brillouin scattering is another technique that has become a standard method for obtaining high-frequency sonic relaxation information (Patterson 1980). It has been used to study a range of polymers and the results correlate well with other relaxation data. Results for polypropyleneglycol illustrate unmerged loss peaks persisting to extremely high frequency (Patterson *et al* 1978).

Photon correlation spectroscopy (Patterson *et al* 1980) is yet another method for observing relaxation. Studies of PMMA and atactic polystyrene near their glass transition regions have yielded results which complement dynamic mechanical studies. The fluctuations accessible are density and orientation based. Frequencies from about 10^{-2} to 10^5 Hz are probed.

Spectroscopic techniques should also become more valuable in relaxation studies owing to the availability of short intense light pulses. One such approach involves the rotational relaxation of organic molecules as measured by photobleaching. A short (picosecond) pulse of polarised light is followed by interrogating pulses that measure the absorption spectra. The first pulse bleaches the solution and the interrogating pulses measure the rate of rotation of molecules into absorbing orientations. In dimethyl ether, pentacene exhibits rotational times of about 90 ps while anthracene relaxes in 17 ps (Huppert *et al* 1980).

Spectroscopic methods are attractive as they can be made to probe specific molecular groups. We expect many variations on the theme.

Finally, it is important to recognise that dielectric techniques are still yielding exciting results. Polyvinylidene fluoride (PVF_2) was demonstrated to be piezoelectric about a decade ago. It is now known to be pyroelectric and, quite recently, it has been shown to possess a ferroelectric phase. Structural studies are being carried out to analyse the complex phase behaviour. Hysteresis loops are observed and the Curie temperature is found to lie near or slightly above the melting point (Furukawa *et al* 1980, Yagi *et al* 1980).

References

Douglass D C and McBrierty V J 1978 *Macromolecules* **11** 766

Furukawa T, Johnson G E, Bair H E, Tajitsu Y and Fukada E 1980 *Ferroelectrics* submitted for publication

Garroway A N, Moniz W B and Resing H A 1978 *Chem. Soc. Faraday Symp.* No 13

Hoch M J R, Bovey F A, Davis D D, Douglass D C, Falcone D R, McCall D W and Slichter W P 1971 *Macromolecules* 4 712

Hoffman J D, Williams G and Passaglia E 1966 *J. Polym. Sci.* **C14** 173

Huppert D, Douglass D C and Rentzepis P M 1980 *J. Chem. Phys.* **72** 2841

McBrierty V J and Douglass D C 1980 *Phys. Rep.* **63** 62

McBrierty V J, McCall D W, Douglass D C and Falcone D R 1970 *J. Chem. Phys.* **52** 512

McCall D W 1966 *J. Phys. Chem.* **70** 949

— 1971 *Acc. Chem. Res.* 223

McCall D W and Douglass D C 1965 *Appl. Phys. Lett.* 7 12

McCall D W, Douglass D C and Falcone D R 1967 *J. Phys. Chem.* **71** 998

Paterson G D 1980 *Crit. Rev. Solid State Mater. Sci.* 9 373

Patterson G D, Jarry J P and Lindsey C P 1980 *Macromolecules* **13** 668

Patterson G D, Douglass D C and Latham J P 1978 *Macromolecules* **11** 263

Schaefer J, Stejskal E O and Buchdahl R 1977 *Macromolecules* **10** 384

Yagi T, Tatemoto M and Sako J 1980 *Polym. J.* **12** 209

Inst. Phys. Conf. Ser. No. 58
Invited paper presented at Physics of Dielectric Solids, 8–11 September 1980, Canterbury

Piezoelectricity and pyroelectricity in poly(vinylidene fluoride)

G R Davies
Department of Physics, The University of Leeds, Leeds LS2 9JT, England

Abstract. There has been a rapid growth in interest in piezoelectricity and pyroelectricity in synthetic polymers in the last decade, following the discovery of such effects in poly(vinylidene fluoride) (PVDF). This material can readily be made into films suitable for the fabrication of a variety of devices such as microphones, headphones, ultrasonic transducers and heat sensors.

It has proved difficult to determine the origins of the effects, since PVDF is a semicrystalline polymer with at least four different crystalline forms, some of which have only recently been discovered and whose structures are not known in detail. At least two of these forms are non-centrosymmetric and it is therefore possible that the observed piezoelectric and pyroelectric effects arise from these crystalline forms. An alternative explanation is possible, however, which merely requires that the crystalline form be polar. Changes in the dimensions or relative permittivity of the crystalline or amophous phase then lead to a macroscopic response.

1. Introduction

The piezoelectric properties of quartz were discovered in the late nineteenth century, and by the middle of the twentieth century piezoelectric activity had been detected in a wide variety of materials including some polymers (mostly biological). In the last decade, however, there has been a massive growth of research into piezoelectricity and pyroelectricity in synthetic high polymers. Most of this effort is concentrated upon poly(vinylidene fluoride) (PVDF) since this yields the highest response of all synthetic polymers.

PVDF is a semicrystalline thermoplastic polymer which is solvent resistant and chemically inert. Being thermoplastic, it can easily be formed into a wide variety of shapes and, in particular, it can be produced in film form in thicknesses down to 6 μm. When suitably treated the film may exhibit a piezoelectric coefficient $\sim 20\,\mathrm{pC\,N^{-1}}$ and a pyroelectric coefficient of $\sim 30\,\mu\mathrm{C\,m^{-2}\,K^{-1}}$. Such a film can obviously be used in a variety of applications including microphones, headphones, hydrophones, heat sensors and keyboards and many of these applications are on the verge of commercial exploitation. Despite this progress in applications, however, the mechanisms of the piezoelectric and pyroelectric responses are not well understood.

Reviews by Fukada (1971), Hayakawa and Wada (1973), and Wada and Hayakawa (1976) deal with piezoelectricity and pyroelectricity in polymers in general (including PVDF) and a short review by Murayama *et al* (1976) deals specifically with PVDF. It is the aim of this limited presentation to describe the properties of PVDF and to discuss possible mechanisms for its piezoelectric and pyroelectric activity. Where possible, reference will be made to more recent work not covered by earlier reviews.

2. General considerations

PVDF is a long-chain polymer with the chemical repeat unit CH_2CF_2. It is about 50% crystalline with a density of ~ 1.76–$1.80\ \mathrm{Mg\,m^{-3}}$. The crystalline melting point is in the region 430–470 K, depending upon the crystal form present and also upon the size and perfection of the crystallites. Typical crystallite dimensions are 10–100 nm perpendicular to the polymer chains and ~ 10 nm in the chain direction. The amorphous regions are 'rubbery' at room temperature, the glass transition being about 230 K. For many purposes, therefore, it is instructive to consider the material as a composite composed of stiff crystalline lamellae in a soft amorphous matrix.

In order to make a virgin sample exhibit piezoelectric or pyroelectric properties it is necessary to 'pole' it (i.e. to subject it to a large electric field of the order of $100\ \mathrm{MV\,m^{-1}}$, usually at a temperature around 370 K, though poling can be achieved at lower temperatures with larger fields). Poling produces many effects including orienting dipoles in the material, separating ionic impurities and injecting charges from the electrodes. It can also cause transformations between different crystalline forms.

3. Crystalline forms

The relevance of much early work on PVDF has been diminished to a large extent by the gradual discovery of at least four different crystalline forms. These are commonly called I, II, III and II_p or β, α, γ and α_p respectively, in order of discovery. The structures of forms I and II are generally accepted to be as described by Hasegawa *et al* (1972) but Weinhold *et al* (1979) reject the structure described by Hasegawa for form III, though they have not yet published a full structure determination for this form. Form II_p is a polar modification of form II suggested by Davis *et al* (1978) and again a full structure determination has not yet been performed. In these structures the covalent bond lengths are reasonably constant and different conformations of the chains are produced by simple rotations about the C–C bonds in the chain 'backbone'. A planar zigzag chain (like polyethylene) is produced by a sequence of trans (T) bonds, and a rotation of 120° about a trans bond converts it to a gauche (G) bond. A similar rotation in the opposite direction produces a gauche′ (G′) bond. Using this notation we may describe the different forms as follows.

Form I. This is monoclinic with $\beta = 90°$, $a = 858$ pm, $b = 491$ pm, and $c = 256$ pm. There are two chains per unit cell with each chain in the fully extended or all-trans configuration. Each chain has a high dipole moment perpendicular to its backbone since all the electronegative fluorine atoms are concentrated on the same side of the chain. These polar chains pack in a parallel array giving a highly polar and non-centrosymmetric unit cell.

Form II. This is orthorhombic with $a = 496$ pm, $b = 964$ pm, and $c = 462$ pm. The two chains per unit cell are in a TGTG′ conformation. This leads to shorter and less polar chains which pack in an antipolar array. The unit cell is centrosymmetric, hence this form cannot be piezoelectric or pyroelectric.

Form III. This has the lattice dimensions $a = 497$ pm, $b = 966$ pm, and $c = 918$ pm. Banik *et al* (1979) have suggested on theoretical grounds that the chain conformation could be TGTG′TG′TG or TTTGTTTG′, either of which is polar. Systematic absences in the x-ray data suggest a glide plane such that two polar chains would have to pack in a polar array. The unit cell would then be polar and non-centrosymmetric.

Form II_p. This is a modification of form II produced by the application of a large electric field ($\sim 100\,\mathrm{MV\,m^{-1}}$). It consists of TGTG′ chains which are packed in a polar array and can be produced by a 180° rotation of half the chains in the form II crystal. The unit cell is polar and non-centrosymmetric with almost identical dimensions to those of form II. The polarisation is about 4/7 of that of form I.

It has been suggested by Servet and Rault (1979) that a fifth form may be produced by suitable annealing of form II_p. The suggested chain conformation is TTTGTTTG′, however, and it is possible that form III has been produced by this process. Indeed, it should be noted that the lattice dimensions for forms II and II_p are virtually identical and that form III has closely similar a and b dimensions to form II with almost exactly double the c dimensions. This means that diffracted beams from the different phases fall in very similar positions and detailed inspection of the relative intensities of the different reflections is required to analyse the relative proportions of the different crystalline forms. It is highly desirable that x-ray analysis of samples should be supported by other techniques such as infrared analysis.

When PVDF is crystallised from the melt it crystallises in form II. It is usual, however, to draw samples either uniaxially or biaxially to a draw ratio of ~4:1 to increase their piezoelectric activity. Since form I chains are more extended than those of form II, this process converts form II to form I, the degree of conversion being greater for higher draw ratios and lower draw temperatures. As relatively few piezoelectric or pyroelectric studies have been performed on samples containing form III, the rest of this paper will concentrate upon samples initially in form I or form II.

The general properties of PVDF will be illustrated by reference to three types of sample. These are isotropic form II, uniaxially oriented form II and uniaxially oriented form I. The isotropic form II samples were 'non-oriented' 50 μm thick capacitor-grade film supplied by Kureha Chemical Industries Co. Ltd, and the oriented form II samples were produced by drawing this film to a 4:1 draw ratio at 160°C and then annealing at constant length for six hours at 120°C. Oriented form I samples were prepared by drawing the isotropic film to a 3.5:1 draw ratio at 55°C then annealing overnight at constant length at 150°C. Work at Leeds has concentrated upon a full characterisation of the properties of these samples and the data which follow are taken mainly from the work of Rushworth (1977).

4. Mechanical and dielectric properties

The properties of PVDF are temperature and frequency dependent. In particular, there are several relaxation processes which lead to high levels of mechanical and dielectric loss. Figure 1 shows typical loci of loss maxima in the $(\log f) - T^{-1}$ plane. The curving line near the centre of the figure indicates the glass transition process in the amorphous regions. Below room temperature this process lowers in peak magnitude but broadens

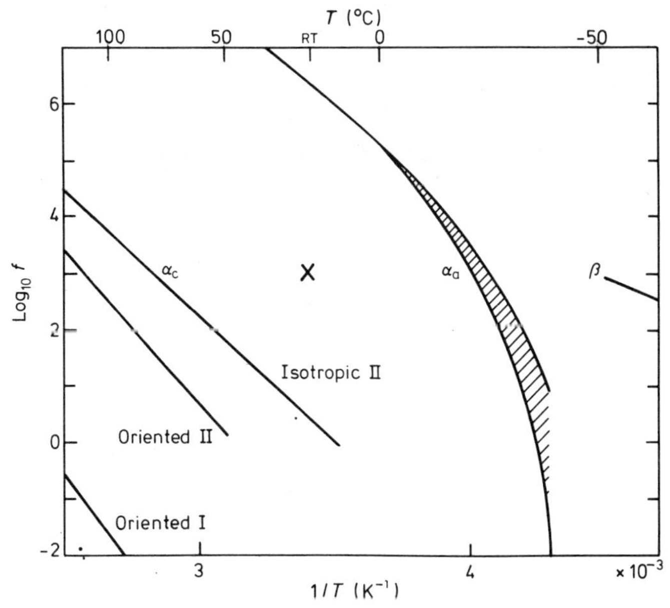

Figure 1. Typical loci of loss maxima for PVDF films.

rapidly in frequency space. The straight lines to the left of the figure represent various crystalline loss processes. It is fortunate that room temperature audio-frequency applications (indicated by the cross) lie in a loss minimum between the glass transition process and the crystalline loss processes. On going to higher frequencies, however, rising levels of loss are encountered due to the glass transition process.

Drawn films are anisotropic and a variety of measurements are required to characterise them fully. We have measured the extensional compliance at different angles to the draw direction and the relative permittivity perpendicular to the plane of the film. Details of typical sample configurations and directions of axes are given in figure 2, and results for the extensional compliance at 1.69 Hz are shown in figure 3.

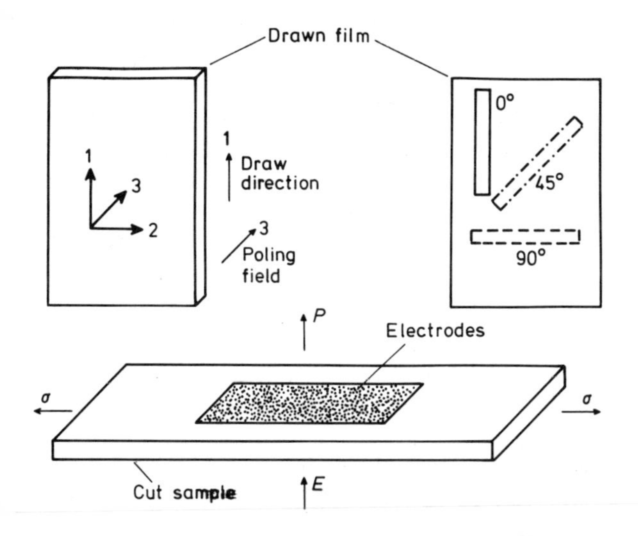

Figure 2. Orientation of axes and samples cut from drawn films.

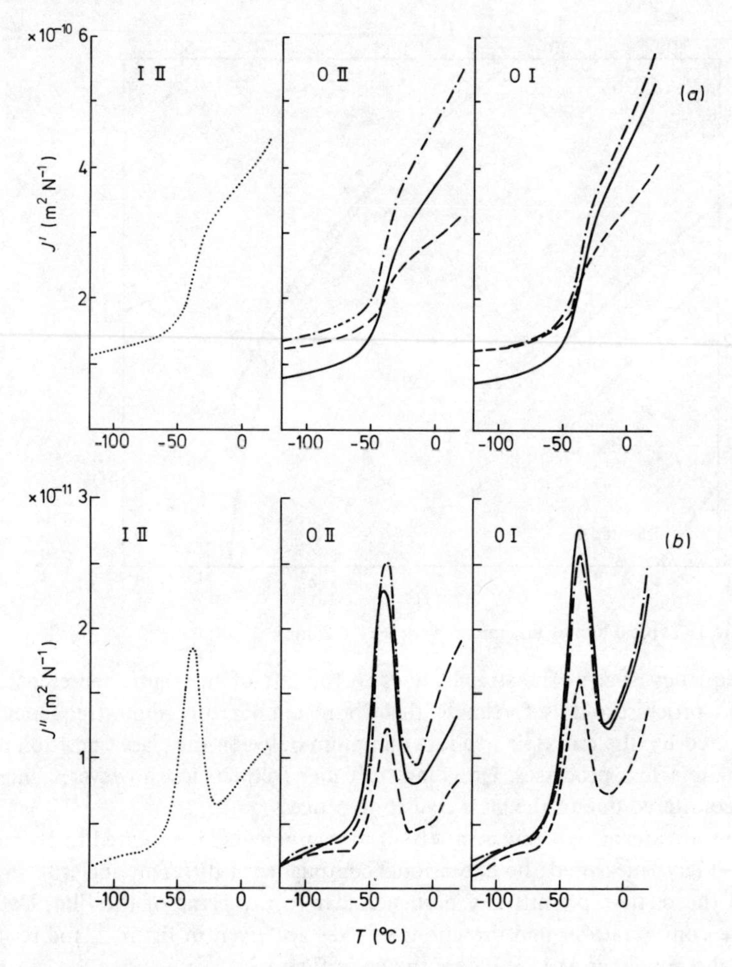

Figure 3. (*a*) The real part (J'), and (*b*) the imaginary part (J'') of the complex compliance of PVDF films measured at 1.69 Hz. I II, isotropic II; O II, oriented II; O I, oriented I; ——, 0°; –·–·–, 45°; -----, 90°.

The main step in these data is due to the glass transition process in the amorphous regions of the sample. It can be seen that the loss due to this process falls away as room temperature is approached and a second loss process is encountered at higher temperatures. It is interesting to note that these samples are stiffest at right angles to the draw direction and softest at 45° to it. This anisotropy of compliance is typical of an interlamellar shear process as described by Davies *et al* (1972) for polyethylene. The high compliance at 45° to the draw direction is due essentially to the tendency of oriented and annealed polymers to form stacks of lamellar crystallites oriented with their large faces perpendicular to the draw direction and separated by amorphous material. A tensile stress applied at 45° to the draw direction therefore produces a large deflection by shearing the relatively soft interlamellar (amorphous) regions.

The permittivity of the samples, also at 1.69 Hz, is shown in figure 4. Note that the drawn samples show a higher room-temperature permittivity than the isotropic sample.

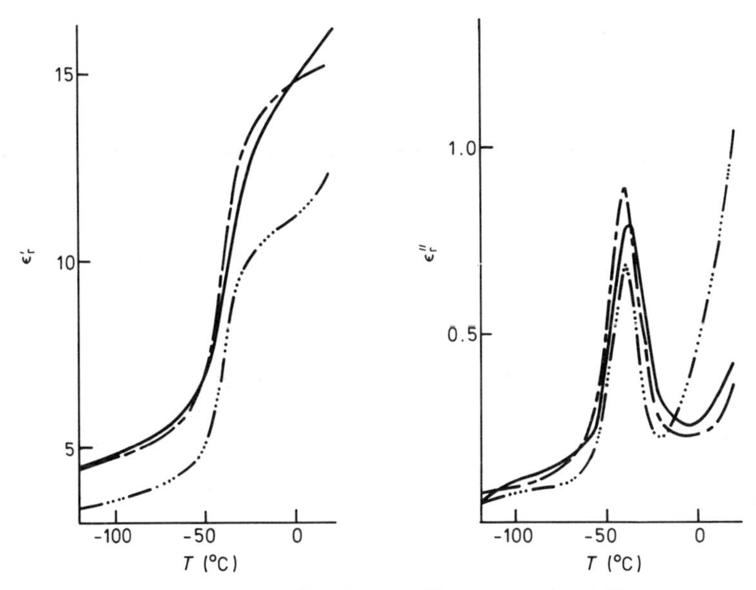

Figure 4. The relative permittivity of PVDF films measured at 1.69 Hz. Isotropic II, —— · · ·; oriented II, —— –; oriented I, ——.

Baird *et al* (1975) attribute this to an orientation of dipole rotation axes in the amorphous phase but Davies and Rushworth (1976) believe it to be an exaggerated 'form factor' effect due to the orientation of the crystal lamellae in the drawn samples. Basically, different mixing laws are required to calculate the bulk permittivity from the permittivities of the two phases. For an isotropic sample, practical experience shows that a logarithmic mixing law is usually appropriate. That is, if we denote the degree of crystallinity by X and the relative permittivities of the bulk, crystalline and amorphous material by ϵ, ϵ_c and ϵ_a respectively, we can write:

$$\log \epsilon = X \log \epsilon_c + (1 - X) \log \epsilon_a \, .$$

In view of the tendency of oriented samples to produce oriented stacks of crystalline and amorphous material, however, it is better to calculate the relative permittivity perpendicular to the stacks by a simple volume-fraction-weighted average of the permittivities of the two phases. In this case, therefore, we use:

$$\epsilon = X\epsilon_c + (1 - X)\epsilon_a \, .$$

Putting $\epsilon_c = 4$, $\epsilon_a = 25$ and $X = 0.5$ into the above equations yields a relative permittivity of 10 for an isotropic sample and 14.5 for an oriented one. Since the assumed values are reasonable for PVDF, it follows that the dielectric properties of bulk samples are strongly influenced by the lamellar orientation.

5. Piezoelectric and pyroelectric properties

As shown by Blevin (1977) or Day *et al* (1974), the magnitude of the piezoelectric and pyroelectric responses increases with poling field, poling temperature and poling time. Day *et al* (1974) have also shown that insufficiently poled samples exhibit a pyroelectric response which is non-uniform across the thickness of the sample and it may be assumed that this is also accompanied by a non-uniformity of the piezoelectric response. The effect is most apparent with thick samples and it is difficult to pole samples greater than one or two hundred microns thick in a uniform manner.

Figures 5*a* and 5*b* show the piezoelectric response (again at 1.69 Hz) for samples 25–50 μm thick with Nichrome gold electrodes poled at 50°C for 30 min at a field of 40 MV m^{-1}. In the matrix notation the data for the 0° samples would be referred to as d_{31} (i.e. the change in polarisation in direction 3 from a change in the tensile stress along direction 1) and that for the 90° samples would be referred to as d_{32}. Since the change in polarisation is usually out of phase with the change in stress, a complex *d* coefficient is defined in an analogous manner to complex permittivity. The quantity d' therefore relates to the change in polarisation in phase with the change in stress, and d'' relates to the out-of-phase response. It can be seen that the form I and form II samples give qualitatively similar results but differ by about a factor of two in their absolute magnitude.

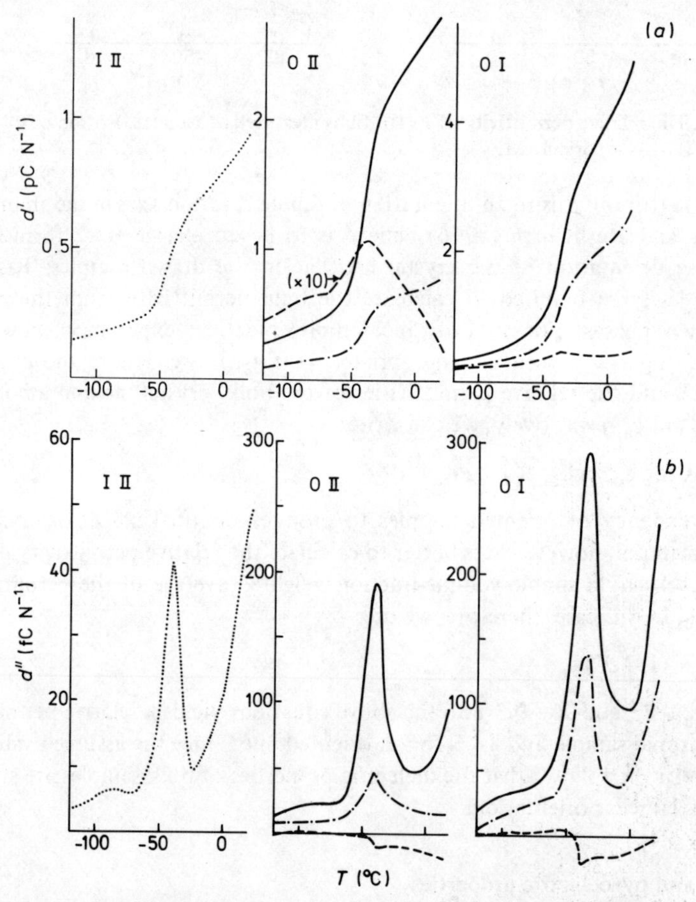

Figure 5. (*a*) The real part (d'), and (*b*) the imaginary part (d'') of the complex piezoelectric strain coefficient of PVDF films measured at 1.69 Hz (samples poled at 40 MV m^{-1} and 50°C for 30 min). I II, isotropic II; O II, oriented II; O I, oriented II; ——, 0°; —·—·—, 45°; - - - - -, 90°.

This similarity in response is repeated in the pyroelectric data shown in figure 6. These samples were poled at 130°C and 100 MV m^{-1} for 30 min and the measurements were made at 20 mHz. It can be seen that all the curves may be roughly superposed by a vertical shift. Since the vertical scale is logarithmic this implies a constant factor difference between the different curves. Again we see that the oriented form I sample yields about twice the response of the oriented form II sample.

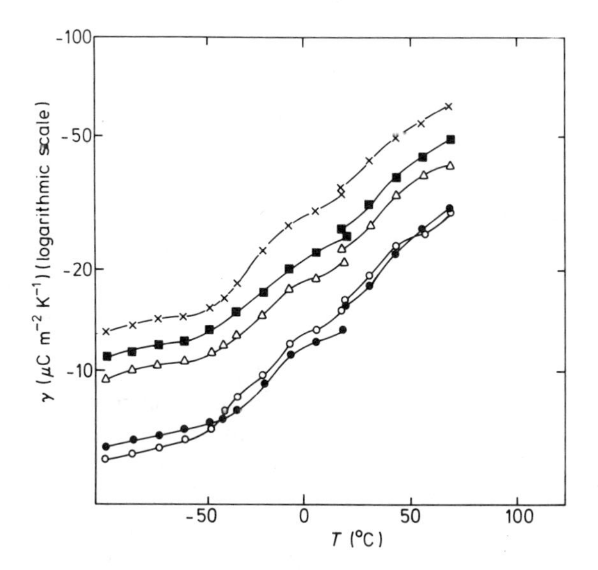

Figure 6. The pyroelectric coefficient (γ_3) of PVDF films measured at 20 mHz (samples poled at 100 MV m^{-1} and 130°C for 30 mins). ×, oriented I; ■, biaxial film; △, biaxial film; ●, oriented II; ○, isotropic II.

It should be emphasised that the description of the crystal form of the sample applies before poling. Davies and Singh (1979) have shown that under these poling conditions a large amount of the oriented form II is converted to form II$_p$ and, to a lesser extent, to form I. A more quantitative estimate is not yet available.

The data in figure 5 are not representative of the maximum response of PVDF but are chosen to show the very similar behaviour in samples containing different crystalline forms. The work of the following authors should be consulted for data covering a wider range of conditions: Ohigashi (1976), Tamura *et al* (1977), Kepler and Anderson (1978) and Klaase and Van Turnhout (1979). As a rough generalisation it may be said that it is relatively easy to produce material with a d_{31} of 20 pC N^{-1} and a pyroelectric coefficient of 30 μC m^{-2} K^{-1} by poling biaxially or uniaxially drawn thin films at about 100 MV m^{-1} and 100°C. It is difficult to measure d_{33} reliably but this is negative and usually greater in absolute magnitude than d_{31}. Table 1, taken from Murayama (1976) gives a comparison of the properties of a typical uniaxially oriented PVDF film with those of other piezoelectric materials. The data quoted for PVDF are appropriate for tensile stresses along the draw direction and polarisation measurements perpendicular to the plane of the film.

6. Mechanisms

It has long been recognised that the piezoelectric activity of PVDF is enhanced by drawing an isotropic film to about 4:1 draw ratio. As mentioned previously, this process tends to convert form II to form I and it was reasonable, therefore, to suppose that the observed

Table 1. Comparison of the properties of PVDF with those of conventional piezoelectric materials. d = d (polarisation)/d (stress), e = d (polarisation/d (strain), g = d (field/d (stress), h = d (field)/d (strain).

Material	Density	Stiffness	ϵ_r	Piezoelectric coefficients				Coupling coefficient
				d	e	g	h	
	($Mg\,m^{-3}$)	(GPa)		($pC\,N^{-1}$)	($mC\,m^{-2}$)	($mV\,m\,N^{-1}$)	($V\,m^{-1}$)	
Quartz	2.65	77	4	2	150	50	390	0.09
PZT	7.5	83	1200	110	9200	10	95	0.31
$BaTiO_3$	5.7	110	1700	78	8600	5	60	0.21
Rochelle salt	1.77	18	350	275	4900	90	340	0.66
PVDF	1.78	3	13	20	60	174	53	0.10

piezoelectric response was due to the intrinsic response of the form I crystals. As has been shown, however, samples initially in form II also display significant activity. Since this form cannot be piezoelectric, this pointed to an alternative explanation in terms of the response of a distribution of trapped charges to an inhomogeneous strain as described in the reviews cited above.

With the discovery of the non-centrosymmetric form II_p in poled form II samples, it is now possible that the response of these samples arises from the genuine response of the form II_p crystals. Indeed, a plausible mechanism is available: if one applies a tensile stress to the form II_p chain in the TGTG′ conformation, it will produce slight rotations about the C–C bonds towards an extended all-trans conformation. This will produce an increase in the dipole moment normal to the chain axis and hence a piezoelectric response. The pyroelectric activity could arise from an increase in the librational motions about the chain axis when the temperature is increased, which would decrease the mean moment perpendicular to the chain axis. Since the form I chain is not perfectly planar, similar mechanisms could apply to the form I crystal as described by Ohigashi (1976).

Difficulties arise, however, when one considers the temperature dependence of the piezoelectric response in the glass transition region. It is reasonable to assume that crystalline properties should not be strongly temperature dependent in this region and that the temperature dependence of the observed piezoelectric response arises from changes in the mechanical and dielectric properties of the amorphous phase. To date, it has not proved possible to construct a model based upon these hypotheses which correctly predicts the observed mechanical, dielectric and piezoelectric properties in this region.

It is my belief that a large portion of the activity arises from the combination of an aligned, polar crystalline phase with a readily deformable amorphous phase. Before developing this argument, however, a short digression is in order.

If we place electrodes of area A on either side of a sample of polarisation P and connect the two electrodes together then the charge q on the plates is given by $q = P/A$. Changes in polarisation can therefore be deduced from the changes in the charges on the plates, if the change in area is known. That is, denoting stress or temperature by x

$$dP/dx = d(q/A)/dx \,.$$

The change in area is difficult to contend with, however, and it has become accepted practice to replace all such differentials by $(dq/dx)/A$ when reporting piezoelectric or pyroelectric coefficients. We therefore wish to calculate changes in the charges on the electrodes in response to changes in temperature or stress, and then apply this latter formula to derive 'experimental' coefficients. With this in mind we can return to the 'polar crystal' model.

Suppose that the crystalline regions between the electrodes contain n dipoles each of moment m perpendicular to the plane of the film. If the film thickness is t then $P = nm/At$ and $q = AP = nm/t$. If the film is uniformly strained and the electrodes deform with the film (a reasonable assumption for evaporated electrodes) then the total number of dipoles between the electrodes does not change and the observed response can arise simply from changes in the thickness of the film.

The 'experimental' quantities to be calculated are of the form:

$$(\mathrm{d}q/\mathrm{d}x)/A = (\mathrm{d}(nm/t)/\mathrm{d}x)/A$$
$$= -(nm/At) \times (\mathrm{d}t/\mathrm{d}x)/t$$
$$= -P \times (\mathrm{d}t/\mathrm{d}x)/t.$$

If we denote stress, expansion coefficient, compliance and Poisson's ratio by σ, α, s and υ respectively then the piezoelectric 'd' coefficients are given by

$$d_{31} = -P \times (\mathrm{d}t/\mathrm{d}\sigma_1)/t = P \times s_{11} \times \upsilon_{31}$$
$$d_{32} = -P \times (\mathrm{d}t/\mathrm{d}\sigma_2)/t = P \times s_{22} \times \upsilon_{32}$$
$$d_{33} = -P \times (\mathrm{d}t/\mathrm{d}\sigma_3)/t = -P \times s_{33}$$

and the pyroelectric coefficient 'γ' is given by

$$\gamma_3 = -P \times (\mathrm{d}t/\mathrm{d}T)/t = -P \times \alpha_3 .$$

The polarisation of the form I unit cell is 130 mC m^{-2} and samples are about 50% crystalline, hence P is about 65 mC m^{-2}; s_{11} and s_{22} are about 0.4 and 0.3 m^2 GN^{-1} as shown by the data for the 0° and 90° samples above, and it is reasonable to assume that $s_{33} = s_{22}$. To the best of my knowledge, the relevant expansion coefficient and Poisson's ratios have not been measured on thin uniaxially oriented films but Booth and Davies (1980) have measured these quantities for a thick uniaxially drawn film containing both form I and form II and find $\alpha_3 = 170 \times 10^{-6}$, $\upsilon_{31} = 0.63$ and $\upsilon_{32} = 0.34$, though this must be regarded as preliminary data.

The predictions of the model based upon the above figures are given in table 2 together with experimental data from Klasse and Van Turnhout (1979) for uniaxially oriented form I and form II samples. The model predictions for form II have been obtained simply

Table 2. A comparison of the predictions of the simple model with the data of Klaase and Van Turnhout (1979).

	γ_3 (μC m^{-2} K^{-1})	d_{31} (pC N^{-1})	d_{32} (pC N^{-1})	d_{33} (pC N^{-1})
Form I (Exptl)	−30	18	1	−27
Form II (Exptl)	−14	10	2	−15
Form I (Model)	−10	16	7	−20
Form II (Model)	−6	9	4	−11

by multiplying the result for form I by 4/7 since this is the ratio of the dipole moments of the form II_p and form I unit cells. This is equivalent to assuming that poling a form II sample produces a sample which is 50% form II_p and that this sample has the same mechanical properties as a form I sample. It can be seen that the model predicts the sign and general magnitude of the effects surprisingly well and there can be no doubt that a large part of the response of PVDF is due to changes in the thickness of the sample.

The model totally neglects the two-phase nature of the sample, however, and consequently cannot include the effects of a change in the permittivity of either phase since the electric field is zero everywhere in a short-circuited homogeneous sample. If we recognise that the sample consists of a crystalline and an amorphous phase then internal fields arise, even in a short-circuited sample, due to the polarisation of the crystalline phase. Inhomogeneous strains also arise due to the differing response of the two phases to changes in stress or temperature. To illustrate this, consider the situation shown in figure 7.

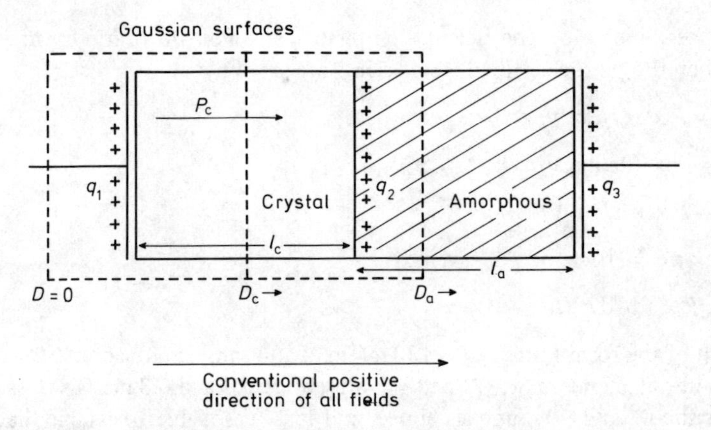

Figure 7. A schematic two-phase system to show how fields arise in a short-circuited sample.

To calculate the displacement D_c in the crystalline phase we consider the Gaussian surface represented by the left-hand box which contains the electrode charge q_1. The displacement outside the electrodes is zero hence

$$D_c = \epsilon_c \epsilon_0 E_c + P_c = q_1/A \ .$$

In the above equation P_c represents the permanent polarisation of the crystalline phase (which may be stress or temperature dependent) and ϵ_c is the relative permittivity of the crystalline phase.

We may calculate the displacement in the amorphous phase in a similar manner by considering the larger Gaussian surface which contains both q_1 and q_2 (this latter charge is included to demonstrate the effects of trapped charges on the crystalline–amorphous interface). Hence

$$D_a = \epsilon_a \epsilon_0 E_a = (q_1 + q_2)/A \ .$$

For short circuit conditions,

$$E_c l_c + E_a l_a = 0 \ .$$

The above equations may be rearranged to show

$$q_1 = \frac{AP_c\epsilon_a l_c - q_2\,\epsilon_c l_a}{\epsilon_a l_c + \epsilon_c l_a}.$$

Differentiation of the above equation with respect to stress or temperature is a straightforward but somewhat lengthy affair. It is readily apparent, however, that changes in the permittivity or dimensions of either phase will result in changes in q_1. Broadhurst *et al* (1978) have fully analysed a two-phase model of this type with a more realistic phase geometry and have attempted to estimate the relative contributions to the piezoelectric and pyroelectric effects from the different mechanisms. Their main conclusions are that the major part of the response comes from changes in dimensions or permittivity and that relatively little comes from changes in the dipole moment of the molecules in the crystalline phase. They also show that q_2 merely serves to reduce the predicted response.

It is probable that the changes in permittivity which lead to an apparent piezoelectric or pyroelectric response are dominated by changes in the amorphous phase and it is instructive therefore to consider the following simple model which illustrates how these changes come about.

Consider the behaviour of a single chain subjected to a tensile force F and an electric field E as shown in figure 8. This chain is composed of n repeat units each of length l and electric moment m, the angle which m makes with l being fixed. For convenience of calculation, the possible orientations of this repeat unit will be restricted to the discrete set shown in the figure. These have been chosen to model the molecular geometry of PVDF which has a permanent dipole moment essentially at right angles to the main chain, and the directions of F and E are appropriate for the calculation of d_{31}.

If we neglect any steric interactions between the repeat units, then the energy (W_i) of a unit in the ith state is given by:

$$W_i = -(mE\cos\theta_i + lF\cos\phi_i)$$

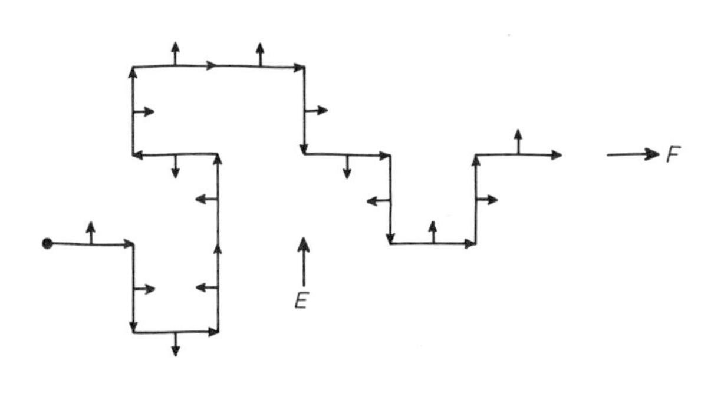

Figure 8. A schematic polar chain to model the mechanical, dielectric and electrostrictive properties of amorphous PVDF (long arrows represent the chain backbone and short arrows represent permanent dipoles; the chain is built from the eight states shown).

where θ_i is the angle between the dipole moment vector and E, and ϕ_i is the angle between the length vector and F. The partition function Z is given by

$$Z = \sum_i \exp(-W_i/kT).$$

The number of repeat units in any state i is then given by

$$n_i/n = \exp(-W_i/kT)/Z.$$

The total moment M in the direction of the electric field E is given by

$$M = \sum_i n_i m \cos\theta_i$$

and the total chain length L in the direction of the force F is given by

$$L = \sum_i n_i l \cos\phi_i \, .$$

The physics of the situation can be clearly seen without any further mathematics. The application of a tensile force F favours the occupancy of states 1 and 2. In the limit of large F, therefore, the electric susceptibility is that of a simple two-state system. In the absence of the tensile force, however, a significant number of units are in states 3, 4, 7 and 8. Since transitions between these states do not involve a change in dipole moment parallel to E, the susceptibility is less than that of the simple two-state system. If there is an electric field present, these differences in susceptibility manifest themselves as a difference in dipole moment which therefore produces an apparent piezoelectric response.

Putting the previous argument on a more mathematical basis, we wish to calculate a general expression for $\mathrm{d}M/\mathrm{d}F$ as a function of F, E and T. Straightforward but rather tedious algebra based on the above equations eventually yields

$$\frac{\mathrm{d}M}{\mathrm{d}F} = \frac{nml}{kTZ^2} \sum_i \sum_{j<i} (\cos\theta_i - \cos\theta_j)(\cos\phi_i - \cos\phi_j) \exp\left[-(W_i + W_j)/kT\right].$$

This is a general expression for any choice of states. If we restrict ourselves to the states shown in figure 8 it reduces to

$$\frac{\mathrm{d}M}{\mathrm{d}F} = 16 \frac{nml}{kTZ^2} \sinh\left(\frac{mE}{kT}\right) \sinh\left(\frac{lF}{kT}\right).$$

We note that both E and F must be finite for this mechanism to yield an apparent piezoelectric response and that this is positive as found for d_{31}.

It would appear, therefore, that it is not enough to have a polar crystalline phase to produce an electric field in the amorphous phase but that the amorphous phase must also be extended to some extent before the above mechanism can operate. Partial orientation will occur naturally where the chains enter the amorphous phase from the crystalline phase and it may therefore be necessary to consider samples as composed of a crystalline phase, an amorphous phase and an interface region, when modelling the properties of a semicrystalline sample.

The above treatment is, of course, incomplete in that it only deals with the properties of a single chain and a quantitative prediction for the *d* coefficient of the bulk sample cannot therefore be made.

7. Conclusions

With the discovery of the polar form II_p there is now emerging a general consensus of opinion that a large part of the piezoelectric and pyroelectric response of PVDF arises from the presence of a polar crystal form coupled with changes in the thickness of the sample. Changes in the relative permittivity of either phase with temperature or stress can also cause a significant response, though it is probable that this occurs mainly in the amorphous phase.

References

Baird M E, Blackburn P E and Delf B W 1975 *J. Mater. Sci.* **10** 1248–51
Banik N C, Taylor P L, Tripathy S K and Hopfinger A J 1979 *Macromolecules* **12** 1015–6
Blevin W R 1977 *Appl. Phys. Lett.* **31** 6–8
Booth M C and Davis G R 1980 unpublished
Broadhurst M G, Davis G T, McKinney J E and Collins R E 1978 *J. Appl. Phys.* **49** 4992–7
Davies G R, Owen A J, Ward I M and Gupta V B 1972 *J. Macromol. Sci. – Phys.* **B6** 215–28
Davies G R and Rushworth A 1976 *J. Mater. Sci.* **11** 782–3
Davies G R and Singh H 1979 *Polymer* **20** 772–4
Davis G T, McKinney J E, Broadhurst M G and Roth S C 1978 *J. Appl. Phys.* **49** 4998–5002
Day G W, Hamilton C A, Peterson R L, Phelan R J Jr and Mullen L O 1974 *Appl. Phys. Lett.* **24** 456–8
Fukada E 1971 *Prog. Polym. Sci. Japan* **2** 329–72
Hasegawa R, Takahashi Y, Chatani Y and Tadokoro H 1972 *Polym. J.* **3** 600–10
Hayakawa R and Wada Y 1973 *Adv. Polym. Sci* **11** 1–55
Kepler R G and Anderson R A 1978 *J. Appl. Phys.* **49** 4490–4
Klaase P T A and Van Turnhout J 1979 *Proc. 3rd Int. Conf. Dielectric Materials, Measurements and Applications* (IEE Conf. Pub. 177) pp411–4
Murayama N, Nakamura K, Obara H and Segawa M 1976 *Ultrasonics* **14** 15–23
Ohigashi H 1976 *J. Appl. Phys.* **47** 949–55
Rushworth A 1977 *Phd Thesis* University of Leeds
Servet B and Rault J 1979 *J. Physique* **40** 1145–8
Tamura M, Hagiwara S, Matsumoto S and Ono N 1977 *J. Appl. Phys.* **48** 513–21
Wada Y and Hayakawa R 1976 *Jap. J. Appl. Phys.* **15** 2041–57
Weinhold S, Litt M H and Lando J B 1979 *J. Polym. Sci. (Polym. Lett.)* **17** 585–9

Inst. Phys. Conf. Ser. No. 58
Invited paper presented at Physics of Dielectric Solids, 8–11 September 1980, Canterbury

Low-temperature effects

W A Phillips
Cavendish Laboratory, Madingley Road, Cambridge CB3 OH3

1. Introduction

It is impossible in a review of this length to give an overall view of low-temperature effects in dielectrics, and so in order to allow the basic physical ideas to be presented the paper will concentrate on measurements of dielectric loss and dielectric constant in two dielectrics. In both cases the effects are primarily the result of one particular impurity, the hydroxyl group, which means that the moment of inertia and the dipole moment are well defined. The experiments cover the frequency range 1 Hz to 11 GHz, and the temperature range 3 mK to 100 K. This temperature range corresponds to energies (i.e. kT) varying from 3×10^{-7}eV to 10^{-2}eV.

Three specific points can be made in connection with low-temperature experiments;

(a) Any description must be quantum mechanical instead of classical. Often, a description of the same microscopic motion involves a different set of parameters at low temperatures than it does at room temperature, so that measurements tend to be complementary. Alternatively, a comparison of measurements at low and high temperatures provides a check on the accuracy of the microscopic description.
(b) The nature of the molecular motion contributing to loss at low temperatures can often be described in simple terms. This is particularly important in amorphous materials where at high temperatures the macroscopic rearrangements involved in the glass transition make an individual-particle model hard to justify.
(c) Novel experiments, sometimes providing very detailed information, are possible at low temperatures. For example, it is relatively easy to apply an electric field sufficiently large to produce saturation, and the use of pulsed high-frequency electric fields leads to the so-called electric echo phenomena.

The remainder of this paper is divided into four sections. An introduction to the simple quantum theory of relaxation processes is followed by two sections describing the behaviour of substituted phenols (similar to commercial antioxidants) in hydrocarbon polymers, and hydroxyl groups in vitreous silica. Finally, a section is devoted to novel low-temperature experiments.

2. Basic physical ideas

The basic ideas can best be illustrated by describing a particular example, a hydroxyl group in a rigid two-fold symmetric potential. This potential is illustrated in figure 1. Because the moment of inertia of the OH group is known, the only variable is the height

0305-2346/81/0058-0064 $01.50

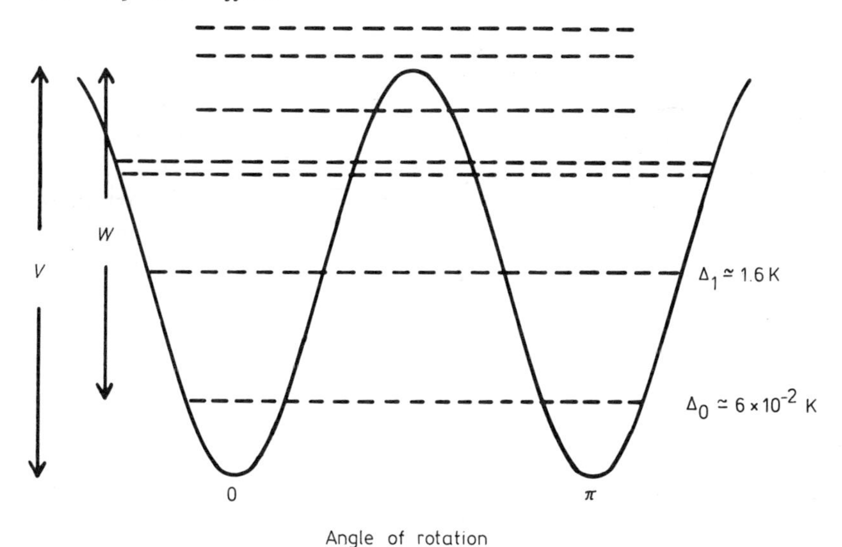

Figure 1. The energy levels given by Matthieu's equation for the rotation of an OH group in a two-fold symmetric potential with a barrier height V of 1000 K. The activation energy W is also marked on the figure, as are the values of the tunnel splitting Δ_0 and Δ_1 for the two lowest energy levels.

of the potential barrier. There are two separate parts to the problem, the calculation of the equilibrium properties and the calculation of the relaxation time.

The allowed energy levels are obtained by solving Schrödinger's equation for a particle moving in a two-fold symmetric potential, equivalent to solving Matthieu's equation. These solutions, giving the allowed energy levels, are shown to scale in figure 1. However, this figure does not show one very important feature, the splitting of the low-lying energy states labelled 0 and 1. The lowest state, 0, is a doublet with a separation Δ_0 equal to 0.06 K, and at low temperatures (say 1 K) it is only this doublet that is significantly occupied. This simplification justifies the use of *two-level systems,* as they are usually called, in interpreting low-temperature properties.

The tunnel splitting Δ_0 is related to the height of the barrier through the approximate formula (Abramowitz and Stegun 1965)

$$\Delta_0' \simeq 9V'^{3/4} \exp(-2V'^{1/2}) \tag{1}$$

where V' is the barrier height V expressed in terms of the energy $\hbar^2/2I$, where I is the moment of inertia, and similarly for Δ'_0.

An important difference between the classical and quantum descriptions must be pointed out. Classically the proton is in either one or the other of the two wells (if the energy is small) but in a quantum description this is not so: in either of the two lowest states the proton is delocalised, and cannot be said to be in a particular well. This has important consequences, the dipole moment for example is zero, and raises the difficult question of principle concerning the transition from a quantum to a classical description. However, in practice this complication only arises for an exactly symmetric potential. A difference in energy between the two wells which is greater than Δ_0 is sufficient to localise the particle. In any amorphous solid such differences are inevitable, so that attention must always be focused on the asymmetric well. However, the extent of the asymmetry is small. As the asymmetry increases the energy difference between the two states increases, and for asymmetries greater than Δ_0 this difference is almost equal to the asymmetry. At a temperature T the asymmetry cannot therefore be much greater than kT: figure 1 could represent an *asymmetric* well important at 1 K.

The dynamical behaviour is very different at high and low temperatures. Classically, the particle moves by thermal activation from one well to the other. In terms of figure 1 the description is very similar. The particle makes a transition from one of the two lowest states to one of the numerous states close to the top of the barrier. This is followed by transitions between these states, and then by a transition down to the other low-lying state. Both descriptions give at high temperatures a relaxation time τ of the form

$$\tau = \tau_0 \, e^{W/kT} \tag{2}$$

where W, the activation energy, is illustrated in figure 1.

At low temperatures the states at the top of the barrier are not involved. The proton passes directly from one low-lying *state* to the other, with the difference in energy taken up by the lattice vibrations. In the asymmetric well the proton is partly localised and the process resembles quantum mechanical tunnelling; for this reason it is known as phonon-assisted tunnelling. The relaxation rate depends on Δ_0^2 (Phillips 1970) and so is again an exponential function of the barrier height, but is an almost linear and not an exponential function of temperature. The overall temperature dependence of the relaxation rate is therefore linear at low temperatures and exponential at high, as shown in figure 2 for values of parameters appropriate to OH in vitreous silica.

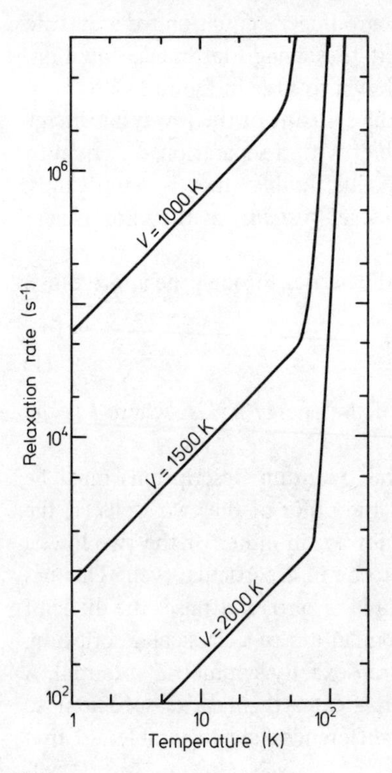

Figure 2. The temperature dependence of the relaxation rate as a function of temperature for three different values of the barrier. The parameters are appropriate to OH in vitreous silica.

Two final points must be made. The first is that phonon-assisted tunnelling is much more sensitive than classical activation to the details of the interaction between the particle involved and the surrounding medium. In order to calculate the relaxation rate the coupling constant must be known. Secondly, as mentioned earlier, the local environment will vary from place to place in the amorphous solid, and may lead to wide variations in the barrier height V. Any such variation will of course be reflected in a very broad (because of the exponentials in equations 1 and 2) distribution of relaxation times. Relaxation peaks may therefore be very broad in an amorphous solid, and may even not show up as true peaks. Section 4 describes such an example but the more straightforward case of a phenol in an inert matrix will be presented first in order to illustrate what happens when the barrier is reasonably well defined.

3. Phenols in hydrocarbon polymers

Figure 3 shows the dielectric loss tangent, tan δ, as a function of frequency at various temperatures between 3.5 mK and 1 K of a dilute solution of 4-methyl-2,6-ditertbutyl phenol in solid poly(4-methyl-1-pentene); it is taken from Isnard *et al* (1980). The hydrocarbon polymer acts essentially as an inert matrix, and interest in this particular combination originated in the fact that this phenol closely resembles commercial antioxidants. Figure 3 is typical of a range of similar phenols in dilute solution, and it is clear that the loss peak arises from the hydroxy proton. For example, the area under the curve, the relaxation strength, is proportional to phenol concentration, and the replacement of the proton by a deuteron gives the expected shift of the relaxation peak to lower frequencies.

The first point to notice is that the peak is relatively well defined, with an area about twice that of a single relaxation time curve of the same height. From the previous discussion, this narrow range of relaxation times implies through equation 1 that the potential seen by the proton varies from one phenol molecule to another by only 15%. It is clear that the reason for this is that the two-fold potential is determined by the phenol molecule itself, with the large t-butyl groups serving to shield the OH from outside influences.

The temperature dependence of the relaxation time, determined from the frequency of maximum loss, resembles that shown in figure 2. Measurements up to 70 K do not

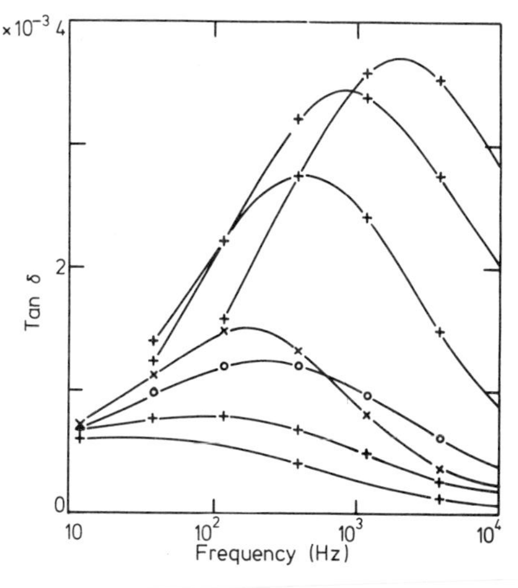

Figure 3. Tan δ in milliradians for a dilute solution of 4-methyl-2,6-ditertbutyl phenol in a 130 μm film of poly(4-methyl-1-pentene) as a function of frequency for different temperatures and measuring voltages. The temperatures are 3,5 mK (lowest curve), 6.5 mK, 11 mK (at two voltages of 25 and 250 mV), 36.5 mK, 222 mK, and 750 mK. The measuring voltage was 100 mV except for the 11 mK curves. (From Isnard *et al* 1980.)

allow the activation energy to be determined, although the results are consistent with the value of approximately 1400 K (1000 cm^{-1}) obtained by Gough and Price (1965) from measurements on 2,4,6-tritertbutyl phenol in decalin solution. It should be mentioned that in certain cases (Isnard *et al* 1980) a slower temperature dependence of the relaxation time is observed between 10 and 50 K, indicating that the simple theory presented in section 2 is not always adequate.

The variation of the area under the loss curve as a function of temperature provides the most detailed information. Between 1 and 50 K the maximum loss (proportional to the area) increases by only 10%. Both classical and quantum theories predict a relaxation strength or area of the form $np^2/3\epsilon_0 kT$, where p is the dipole moment and n is the effective number of contributing dipoles. The local field factor should be small for a hydrocarbon matrix, and has been neglected. A constant relaxation strength implies an effective number of dipoles increasing as T. The obvious interpretation is that there exists as expected an almost constant distribution of asymmetry energies: only those molecules for which the difference between the two lowest energy levels is of order kT or less will contribute to the loss, and the number of such molecules is proportional to kT. This cancellation between the effective number of dipoles and the individual relaxation strengths is to be expected whenever the spread in asymmetry energies is considerably greater than Δ_0, and the temperature range over which it occurs provides a measure of this spread.

Below 1 K the decrease in relaxation strength arises because Δ_0 is large in comparison with kT. Any relaxation process depends on the fact that the equilibrium of the system changes when an electric field is applied. If $\Delta_0 \gg kT$ then thermal equilibrium both before and after application of the field requires all protons to be in the lowest energy state, and the relaxation strength is zero. For a single value of Δ_0 the relaxation strength falls off as $\text{sech}^2\, \Delta_0/kT$ (Frossati and Gilchrist 1977), as shown by the broken curve in figure 4. If

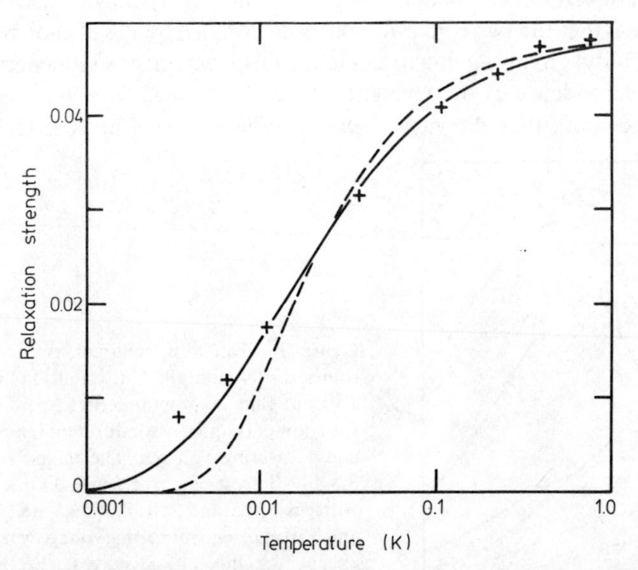

Figure 4. Relaxation strength for the results of figure 3 as a function of temperature. The broken curve is for a single value of the tunnel splitting of 16 mK and the full curve is for a distribution of Δ_0. The crosses are experimental values.

the loss curve is broader than that corresponding to a single relaxation time a suitable average over Δ_0 must be performed. Such an average is also shown in figure 4, where the full curve is calculated using the distribution of relaxation times needed to fit the loss curves of figure 3, and gives good agreement with experiment. The mean value of $\Delta_0/2$ needed to fit the data is 14 mK (290 MHz) which from equation (1) gives a mean barrier height of about 1400 K. The variation in barrier heights mentioned earlier of 15% was derived from this distribution of Δ_0.

These low-temperature experiments give results that are consistent with room temperature measurements. The value of the barrier height agrees reasonably well with the activation energy derived by Gough and Price (1965), and although no direct measurements of Δ_0 are available in the substituted phenols, the value of 290 MHz is consistent with the value of 112 MHz found for pure phenol in the vapour phase (Kojima 1960), bearing in mind the slightly larger barrier found in phenol.

In summary, these low-temperature measurements confirm that a simple rigid rotator model is adequate to explain the motion of the OH group in phenols, and are a useful complementary method of obtaining information on local potentials.

4. OH in vitreous silica

There are a considerable number of dielectric and other measurements that compare 'water-free' and 'wet' vitreous silica. The chemistry of water in silica has been studied in connection with the growth of oxide films on silicon (Doremus 1976) and with the optical absorption arising from OH stretching vibrations, of particular importance for optical fibre applications (Walrafen 1974). The general conclusion of these studies is that OH groups are chemically bonded to the silica network, but that the local environment varies greatly from one OH group to the next, giving a broad absorption line. Because of these strong interactions each OH can be considered to move in an asymmetric double potential, with a wide variation in barrier height from one to another.

The contribution of OH to tan δ is shown in figure 5 (Phillips 1980), obtained by subtracting tan δ for Suprasil W (OH free) from that of Suprasil I (1200 PPM OH). It is less satisfactory to work in terms of temperature as the variable instead of frequency, but

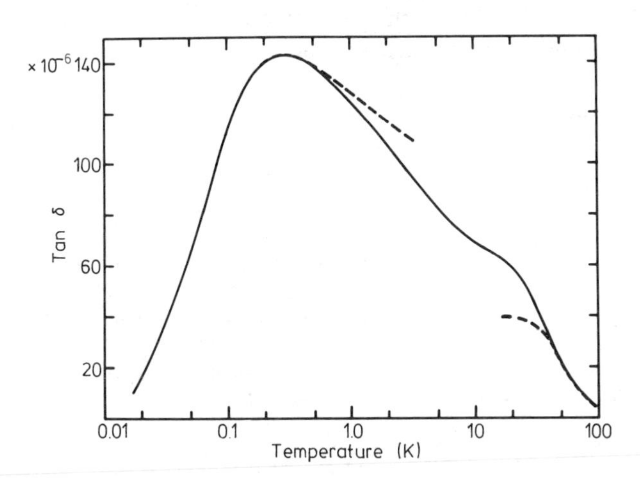

Figure 5. The dielectric loss arising from the presence of 1200 PPM of OH in vitreous silica. The full curve, which is the difference between samples containing 1200 PPM and essentially zero OH, agrees with the classical calculation at high temperatures and with the quantum calculation at low temperatures. In the intermediate temperature range neither theory agrees with experiment, as shown by the dashed continuations.

in this case the wide range of relaxation times leaves little alternative. However, the extremely small loss above 100 K means that this range of relaxation times must be limited. Equation (2), the classical result, will be valid in this temperature range, and implies in terms of the barrier picture of figure 1 that there exist very few high barriers. Here, therefore, as opposed to the case treated in section 3 there is a broad distribution of energy barriers which falls off at high values.

The argument can be put on a quantitative basis by using a specific form for the distribution of energy barriers. An exponential distribution of the form $1/W_0 \exp(-W/W_0)$ combined with equation (2) and the Debye equations gives a good fit to the results above 50 K, as shown in figure 5 (for details of the analysis see Gilroy and Phillips (1980)). This distribution incorporates the high-barrier cut-off in a natural way, but is clearly not the only choice. However, a gaussian form for the distribution does not in practice give as good a fit.

At low temperatures the important parameter is Δ_0, which can be calculated from V using equation (1). The distribution function for Δ_0 can therefore be derived from that for W, relating V to W as shown in figure 1. This distribution function for Δ_0 could then be used to calculate the low-temperature dielectric loss. However, the quantum theory has already been carried through by Frossati *et al* (1977) using an *assumed* form for the distribution of Δ_0. The form they used is almost identical to that calculated from the classical limit, showing that the two calculations are completely consistent.

The calculation can be carried further to provide a consistent picture of *all* the low-temperature experiments on OH in silica. For example, measurements of the real part of the dielectric constant, ϵ_r, between 100 Hz and 11 GHz give values for the variation of ϵ_r with temperature (Frossati *et al* 1977, von Schickfus and Hunklinger 1976). This variation is determined by direct *resonance* absorption between the two lowest energy levels, and cannot be directly inferred from a Kramers–Kronig transform of the *relaxation* contribution to the loss. However, the variation of ϵ_r with T can be derived from the distribution function for Δ_0, and as shown in figure 6 the predictions of the model are in good agreement with experiment. Other properties can also be calculated; the specific heat derived from the effective number of OH groups is close to the observed difference between Suprasil I and Suprasil W at very low temperatures (Phillips 1980, Lasjaunias *et al* 1975).

Two general points should be made. The first is that the behaviour of OH in silica is very different to that of the substituted phenols, where the barrier is almost the same for

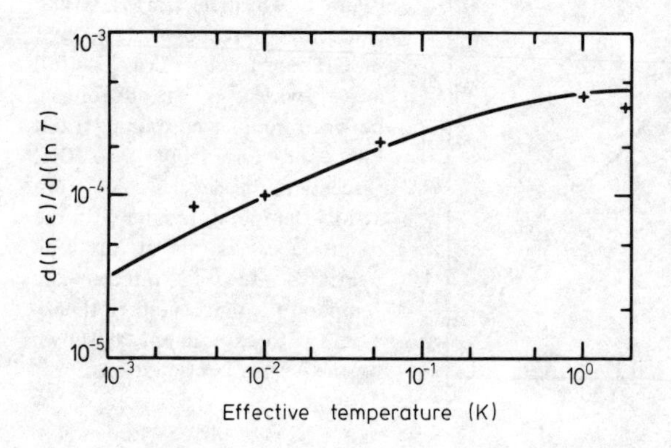

Figure 6. The full curve is the logarithmic slope of ϵ_r in vitreous silica containing 1200 PPM OH, plotted as a function of temperature. The crosses are experimental values, and the curve is normalised to the point at 1 K.

all OH groups. Secondly, a consistent quantitative interpretation can be made using the Debye equations, basic to all these calculations, together with a physically plausible distribution of potential barriers.

5. Low-temperature experiments

A number of experiments are possible only at low temperatures. The first is a matter of technique; the calorimetric method (Vincett 1969) allows the accurate measurement of low losses on small samples over a wide frequency range, but is convenient only up to about 10 K. Secondly it is possible at low temperatures to apply electric fields that are sufficiently large to satisfy the condition $pE \sim kT$. Field-dependent effects are illustrated in figure 3, although the interpretation in this particular case is not straightforward, as the fields involved appear to be too small to produce saturation. However, in general the application of large fields can give a direct measure of the dipole moment (Phillips 1970).

Undoubtedly the most interesting low-temperature dielectric experiment is the observation of electric echoes. The experiment can be described simply, although the explanation is not obvious to those unfamiliar with the corresponding magnetic spin-echo effect. A sample of Suprasil I (containing OH) is held at as low a temperature as possible (10 mK) either in a microwave cavity or between the plates of a capacitor. A short pulse (0.5 μs) of radio-frequency power (500 MHz) is applied to the sample, followed by a second pulse, of the same frequency and amplitude but twice the length, about 10 μs later. If the amplitude of the pulses is chosen correctly a spontaneous signal, the echo, is observed 10 μs after the second pulse. The amplitude of the echo, studied as a function of the amplitude and separation of the pulses, gives detailed information on the dipole moments and interactions between dipoles.

The echo amplitude is a maximum if the electric field E satisfies the condition

$$p_i E\tau/\hbar = \pi/2$$

where p_i is the *induced* dipole moment and τ the pulse length. The induced moment is related to the permanent dipole moment, important in the relaxation experiments, through the asymmetry and the tunnel splitting Δ_0. Figure 7, from Golding *et al* (1979) shows the effect on the echo amplitude of varying the amplitude of the electric field in a number of

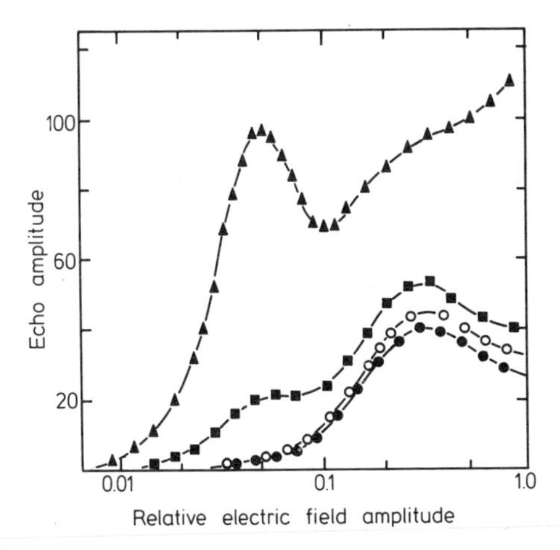

Figure 7. The amplitude of the spontaneous echo in four samples of silica, as a function of electric field strength, taken from Golding *et al* (1979). The lowest two curves (●,○) correspond to samples containing almost no OH, the third (■) contains about 200 PPM, and the upper curve (▲) corresponds to a sample with 1200 PPM OH.

samples of silica containing different concentrations of OH. The effect of OH is obvious, and the experiment also indicates the presence of other dipoles with a much smaller moment.

With a correct choice of experimental conditions the echo amplitude decays exponentially with pulse separation. A relaxation time can be defined from this exponential, but it is not equivalent to the normal relaxation time derived from the relaxation experiments. In magnetic terms it is T_2 rather than T_1 (the normal relaxation time) and, in general, measures the interactions between the dipoles themselves and not the interaction between dipoles and lattice vibrations. More complicated experiments, using three pulses, can be used to measure T_1 but a satisfactory explanation of all these experiments requires theory that is beyond the scope of this brief review (see, for example, the article by Golding and Graebner (1980a)). However, in conclusion it must be emphasised that these techniques should prove extremely valuable for studying electrically active groups in dielectrics, and indeed have already been used to investigate more fully the OH groups in polyethylene (Golding and Graebner 1980b).

References

Abramowitz M and Stegun I A 1965 *Handbook of Mathematical Functions* (New York: Dover)

Doremus R H 1976 *J. Phys. Chem.* **80** 1773–7

Frossati G and Gilchrist J le G 1977 *J. Phys. C: Solid State Phys.* **10** L509–13

Frossati G, Gilchrist J le G, Lasjaunias J C and Meyer W 1977 *J. Phys. C: Solid State Phys.* **10** L515–19

Gilroy K S and Phillips W A 1980 *Phil. Mag.* to be published

Golding B, von Schickfus M, Hunklinger S and Dransfeld K 1979 *Phys. Rev. Lett.* **43** 1817–21

Golding B and Graebner J E 1980a in *Amorphous Solids: Low Temperature Properties* ed. W A Phillips (Berlin: Springer)

— 1980b *Phys. Rev. Lett.* **44** 899–902

Gough S R and Price A H 1965 *Trans. Faraday Soc.* **61** 2435–41

Isnard R, Frossati G, Gilchrist J le G and Godfrin H 1980 *Chem. Phys.* to be published

Kojima T 1960 *J. Phys. Soc. Japan* **15** 284–7

Lasjaunias J C, Ravex A, Vandorpe M and Hunklinger S 1975 *Solid State Commun.* **17** 1045–8

Phillips W A 1970 *Proc. R. Soc.* **A319** 565–81

Phillips W A 1980 *Phil. Mag.* to be published

von Schickfus M and Hunklinger S 1976 *J. Phys. C: Solid State Physics* **9** L439–42

Vincett P S 1969 *J. Phys. D: Appl. Phys.* **2** 699–710

Walrafen G E 1974 *J. Chem. Phys.* **69** 297–8

Inst. Phys. Conf. Ser. No. 58
Invited paper presented at Physics of Dielectric Solids, 8–11 September 1980, Canterbury

Nonlinear optical interactions

J Jerphagnon
Centre National d'Etudes des Telecommunications, 22301 Lannion, France

Abstract. Nonlinear optics encompasses a large variety of phenomena inducing a nonlinear response of matter to electromagnetic radiation. One can for instance mention the electro-optical Pockels and Kerr effects, optical rectification, nth order harmonic generation, two-photon absorption. . . After briefly giving the phenomenological definition of nonlinear optics, we focus attention on second-order susceptibilities. Comments are made on the several approaches to the theoretical description of the origin of nonlinear optical phenomena and their relation to other physical and chemical properties of the material.

A semi-empirical approach aiming at calculating the macroscopic susceptibility from the nonlinear optical properties of microscopic building units (bonds, polyhedra, molecules . . .) is considered in some detail and applied to several families of materials. The microscopic origin of the optical nonlinearities of organic molecular crystals is discussed in comparison with the case of inorganic compounds.

1. Introduction

Although nonlinear optical effects such as the electro-optical Pockels and Kerr effects had been known for some time, nonlinear optics really started to grow with the development of lasers less than 20 years ago. It is a very active field and evidence of new nonlinear optical effects is still being accumulated. A general description of the field and its latest developments is given in Bloembergen (1965, 1977), Zernike and Midwinter (1973) and Rabin and Tang (1975). Considerable progress has been made in the theory of nonlinear interactions in solids as well as in the realisation of nonlinear optical devices. At the same time these devices have provided spectroscopists with new coherent and tunable light sources thus enabling new investigations of solids. The field of nonlinear spectroscopy (Bloembergen 1977) is expanding very rapidly.

A great deal of effort has been devoted to trying to explain the origin of the nonlinearities in solids in order to be able to predict the properties of new kinds of materials. This communication aims to make a brief review of the state of the art in understanding the lowest-order optical nonlinearities of dielectric solids.

2. Second-order optical susceptibilities

Nonlinear phenomena arise from a nonlinear response of a medium to incident radiation. The customary linear constitutive relation between the electric field (**E**) and the electric polarisation (**P**) is no longer valid; it has to be changed to a more complex expression involving higher powers of **E**. In the electric dipole approximation, the constitutive

0305-2346/81/0058-0073 $01.50

relation for an incident radiation with frequency components ω_i, ω_j is as follows

$$\mathbf{P}(\omega_i) = \epsilon_0 \chi^{(1)}(\omega_i) \cdot \mathbf{E}(\omega_i) + \epsilon_0 \chi^{(2)}(-\omega_i; \omega_j; \omega_l) : \mathbf{E}(\omega_j)\mathbf{E}(\omega_l) + \epsilon_0 \chi^{(3)}(-\omega_i; \omega_j; \omega_l; \omega_m) : \mathbf{E}(\omega_j)\mathbf{E}(\omega_l)\mathbf{E}(\omega_m) + \ldots \quad (1)$$

where ϵ_0 is the permittivity of free space and $\chi^{(n)}$ an $n+1$ rank tensor. The larger n, the more difficult it is to detect the associated phenomenon so that intense laser fields are needed for observing sizable effects. $\chi^{(1)}$ is the linear susceptibility related to the dielectric constant ϵ by $\epsilon = 1 + \chi^{(1)}$. $\chi^{(2)}$ is responsible for second harmonic generation ($\omega_j = \omega_l$), optical rectification ($\omega_l = -\omega_i$), linear electro-optic or Pockels effect ($\omega_l \cong 0$), and various three-frequency processes such as difference frequency mixing, sum frequency generation and parametric oscillation. To $\chi^{(3)}$ correspond the third harmonic generation ($\omega_j = \omega_l = \omega_m$), the quadratic electro-optic effect or Kerr effect ($\omega_l = \omega_m \simeq 0$), two-photon absorption ($\omega_j = -\omega_l = \omega_m$, imaginary part), and the self-focusing of light ($\omega_j = -\omega_l = \omega_m$, real part) as well as coherent Raman, Brillouin and Rayleigh scattering (table 1).

Table 1. Low-order nonlinear optical processes.

$\chi^{(2)}$	$(-\omega_i; \omega_j; \omega_l)$	
	$2\omega = \omega + \omega$	Second harmonic generation
	$\omega = \omega + 0$	Linear electro-optics (Pockels effect)
	$0 = \omega - \omega$	Optical rectification
	$\omega_3 = \omega_1 + \omega_2$ $\omega_3 = \omega_1 - \omega_2$	Frequency mixing
$\chi^{(3)}$	$(-\omega_i; \omega_j; \omega_e; \omega_m)$	
	$3\omega = \omega + \omega + \omega$	Third harmonic generation
	$2\omega = \omega + \omega + 0$	DC electric field induced second harmonic generation
	$\omega = \omega + 0 + 0$	Quadratic electro-optics (Kerr effect)
Real part	$\omega = \omega - \omega + \omega$	Self-focusing
Imaginary part	$\omega = \omega - \omega + \omega$	Two-photon absorption
	$\omega_3 = \omega_1 + \omega_1 - \omega_2$	4 waves mixing
Imaginary part	$\omega_2 = \omega_2 + \omega_1 - \omega_1$	Stimulated raman scattering Raman mode $q = \omega_1 - \omega_2$

From now on we shall restrict ourselves to the lowest-order nonlinearity, $\chi^{(2)}$, which results from several mechanisms. The relative contributions depend on the position of the electromagnetic frequencies compared to the electronic and vibrational intrinsic frequencies of the medium. If ω_i, ω_j and ω_l are well below the infrared absorption, the interaction of the electromagnetic fields with the lattice masses provides the major contribution to $\chi^{(2)}$, whereas if ω_i, ω_j and ω_l are below the electronic transitions but well above the lattice absorption, the electromagnetic fields interact with the electrons. If one of the frequencies is below the infrared absorption and the two others above it but below the electronic resonance, then one field interacts with the ionic lattice which drives a change in the electronic states interacting with the other two fields. The linear electro-optic effect **r**, and the closely related optical rectification correspond to this latter situation, while there is the sole electronic contribution, **d**, to $\chi^{(2)}$ for second harmonic generation. It is therefore possible to account for the frequency dispersion of $\chi^{(2)}$ by relating **r** and **d** through the Raman scattering efficiencies of LO and TO phonons of the crystal (Faust and Henry 1966, Kaminow 1967, Johnston 1970) (figure 1).

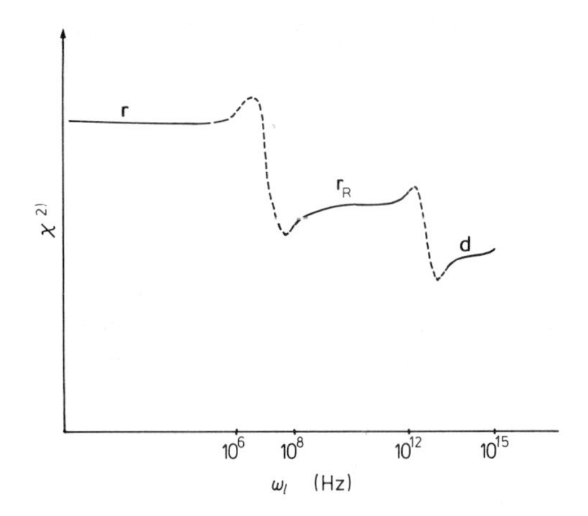

Figure 1. Frequency dispersion of $\chi^{(2)}$. The resonance in the $10^6 - 10^8$ Hz range corresponds to coupling with acoustic vibrational modes, while the resonance around 10^{13} Hz is due to interaction with the Raman-active optical mode.

When all the electromagnetic frequencies fall in the transparency range, $\chi^{(2)}$ satisfies intrinsic symmetry relations (Armstrong *et al* 1962). It has been conjectured by Kleinman (1962) that **d** is frequency independent since the driving mechanism is related to electronic processes. The so-called Kleinman's conditions which state that all the coefficients d_{ijk} of the third rank tensor **d** are real and invariant to any permutation of *ijk* have been shown to be valid in most cases. In addition to the intrinsic symmetries $\chi^{(2)}$ obeys the crystallographic symmetry requirements. In particular $\chi^{(2)} = 0$ if there is a centre of symmetry.

Very different approaches have been used to theoretically determine $\chi^{(2)}$ and **d** in particular. A number of rigorous quantum mechanical treatments have been performed and recently reviewed by Flytzanis (1975). They lead to expressions of the following kind

$$d_{ijk} \propto \sum_{n,n',n''} \int d^3K f_{nK} \frac{p^i_{nn'} \, p^j_{n'n''} \, p^k_{n''n}}{(E_{nn'} - 2\omega)(E_{nn''} - \omega)} \tag{2}$$

where f_{nK} is the Fermi–Dirac function corresponding to the Bloch state $|n, K\rangle$ with energy $E_n(K)$; $p_{nn'} = \langle n, K | \mathrm{p} | n', K \rangle$ where p is the dipole moment operator.

These general *ab initio* quantum mechanical calculations are in fact not very useful in actual evaluations of **d** because they require knowledge of the excited state energies and dipole matrix elements which are generally unknown. In order to be able to predict nonlinear optical properties of new materials one is therefore led to try empirical models. Among them the simplest is the anharmonic oscillator model (Lax *et al* 1962, Garret and Robinson 1966) with a non-parabolic potential V. The main interest of the anharmonic oscillator model is to give some grounds for the remarkable behaviour of Miller's nonlinear coefficient δ defined by Miller (1964)

$$E_i(\omega_3) = \frac{1}{\epsilon_0} \sum_{ijk} \delta_{ijk}(-\omega_3; \omega_1; \omega_2) P_j(\omega_1) P_k(\omega_2). \tag{3}$$

As a consequence the relation between **d** and δ is

$$d_{ijk} = \epsilon_0 \sum_{lmn} \chi_{il}^{(1)}(\omega_3)\chi_{jm}^{(1)}(\omega_2)\chi_{kn}^{(1)}(\omega_2)\delta_{lmn}.$$

While more than four orders of magnitude are necessary to scale the variation of $\chi^{(2)}$ among the nonlinear crystals, the values of δ are nearly the same for all known materials.

3. The molecular engineering approach

Between the sophisticated quantum mechanical calculations and the very simple anharmonic oscillator model there is a large variety of semi-empirical theories; one can consider, for instance, those aiming at predicting the nonlinear optical properties of large classes of materials by considering the physicochemical properties (in particular the chemical bond) in connection with the crystal structure. The concept of decomposing the index of refraction into a sum over contributions of individual bonds has been highly successful (Denbigh 1940). As first suggested by Robinson (1967) the same approach can be used in the case of nonlinear optical susceptibilities. More precisely, the following three-step method allows one to describe quantitatively the microscopic origin of **d**.

1. Decompose the material into an assembly of microscopic units such as diatomic bonds, tetrahedra, octahedra, etc. . . The same microscopic unit can be found in several materials.
2. Describe the microscopic unit nonlinearity, β, by using simplified models with only a few parameters (ionicity, polarisability. . .).
3. Express the macroscopic tensor **d** as the sum of the contributions β_i of the various microscopic units within the crystal

$$\mathbf{d} = \sum_i \beta_i. \qquad (5)$$

Note that when following such a method one has to deal with the difficult problem of local field effects. To account for these effects one can try to actually evaluate local field factors. In an alternative approach the microscopic polarisabilities are calculated in terms of properly chosen macroscopic parameters, providing microscopic polarisabilities somehow renormalised with respect to local field corrections.

The customary cartesian tensor components are not very suitable when changes of the coordinate frames are made. Besides the fact that lengthy calculations are involved, it is almost impossible to follow up to the macroscopic scale all the consequences of the assumptions made at the microscopic level. One has therefore been led to introduce the irreducible tensor formalism (Jerphagnon 1970, Jerphagnon *et al* 1978). It is interesting to note that the decomposition of **d** and β into vector and septor parts (figure 2) strongly suggests simple relations between different physical properties, such as for instance the linear relationship between the vector part of δ and the spontaneous polarisation $\mathbf{P}_s$ (Jerphagnon 1970).

Several models (see Flytzanis 1975 for a review) following the bond polarisability approach have been proposed for binary semiconductors, most of them taking advantage of the recent development of the bond theory (Phillips 1969, 1973). Two parts can be distinguished in the energy gap averaged over the Brillouin zone. The homopolar part

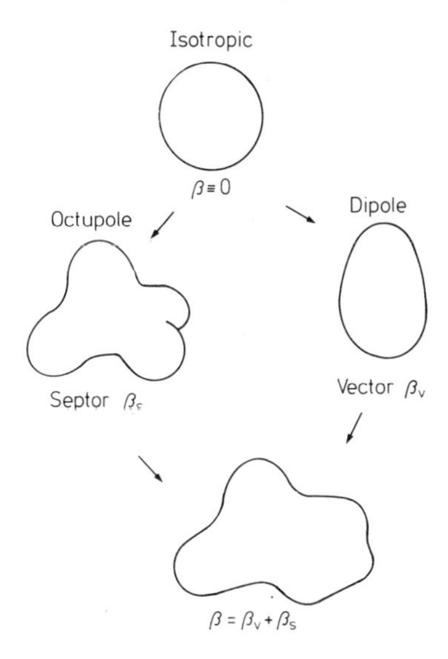

Figure 2. Decomposition of a third-rank tensor into irreducible parts. The vector part corresponds to a dipole moment and the septor part to an octupole moment (pure sp^3 hybridisation).

depends on the total length of the bond and contributes in pure covalent as well as in other binaries. The heteropolar part C is specific of the ionicity of the bond and is related to the valence charge and covalent radii of the atoms. An applied electromagnetic field induces a change in C which results in a nonlinear optical susceptibility β for the bond. In the bond charge model (Levine 1969) the change in C corresponds to a displacement of the very mobile bond charge due to the overlap of the wavefunctions of the atoms of the bond while for Tang (1973) the change in C results from an electric-field-induced charge transfer between the atoms. Both models have been rather successfully applied to ternary semiconductors (Chemla 1971, Scholl and Tang 1973) while Levine (1973) has extended the bond charge model to a number of crystals of various kinds.

Calculations for the niobates and tantalates have been performed by Jeggo and Boyd (1970). They chose for the NbO and TaO bonds a one-dimensional approximation neglecting the polarisability $\beta_\perp$ perpendicular to the direction of the bond compared to the polarisability $\beta_\parallel$ parallel to the bond. Such an hypothesis has been shown to be correct for binary semiconductors (Flytzanis and Ducuing 1968, 1969) but may not hold for other classes of materials. Bergman and Crane (1974) developed with success a three-dimensional model of the chemical bond which they applied to various oxides, in particular the iodates and the oxide systems ABO_x (A = Li, Na, Ba, . . . ; B = Nb, Ti, I . . .).

Wemple and Di Domenico (1968, 1972) had a different phenomenological approach for calculating $\chi^{(2)}$ in materials with tungsten bronze structure ($BaTiO_3$ and related compounds). The oxygen polyhedron is considered as an entity and no assumption on the particular bonds is introduced. The building block of these crystals is the BO_6 octahedron, as shown in particular by the study of the UV absorption spectra (Kurtz 1966). In the ferroelectric phase the energy bands, and consequently the optical susceptibilities, are shifted from their values in the paraelectric phase by the spontaneous polarisation $\mathbf{P}_s$ (figure 3). Using the polarisation potential concept, $\chi^{(2)}$ of the oxygen octahedra ferroelectrics have been related through $\mathbf{P}_s$ to a unique quadratic electro-optic tensor. Satisfactory results are obtained for the linear electro-optic effect, but the model is more questionable for $\mathbf{d}$.

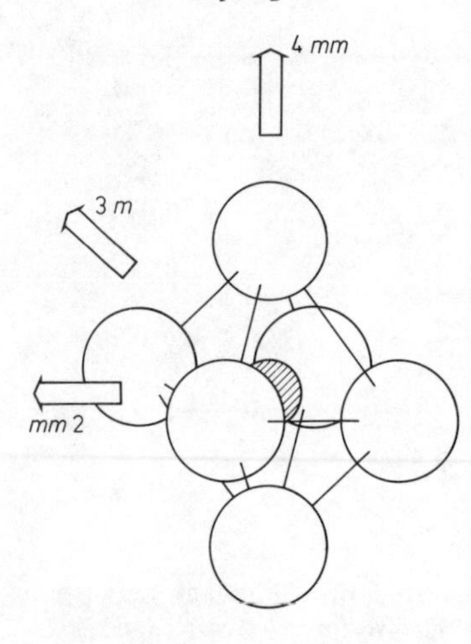

Figure 3. BO_6 octahedron of the tungsten bronze and related structures. The direction of $\mathbf{P}_s$ is shown for the different ferroelectric phases.

In the molecular engineering approach difficulties can be encountered in determining the most suitable building block. There is no question about the choice of the microscopic unit in the case of molecular crystals since in these materials the intermolecular forces are much weaker than the intramolecular ones. Exploratory measurements by second harmonic generation in powders (Kurtz and Perry 1968, Davydov *et al* 1970, Jerphagnon 1971) have shown that large optical nonlinearities are associated with the presence in the molecule of benzene or analogous rings with radicals of different nature (donor, acceptor) asymmetrically substituting for the hydrogen atoms. More recently several systematic studies of families of organic crystals have been performed which corroborated the key role played by delocalised electrons (Chemla *et al* 1979 and references therein). In particular a molecular engineering program has been developed in our laboratory to find new classes of crystals to be used in parametric devices. The study of monosubstituted benzenes showed that the major part of β of the substituted molecule has to be attributed to the distortion of the Π electron system of the molecule M by the interaction with the substituent group R. The so-called equivalent internal field (EIF) model has been proposed which related β_{R-M} to the mesomeric moment $\Delta\mu_R$ of the substituent group according to

$$\beta_{R-M} \propto \left[\frac{\gamma_M}{\alpha_M}\right] \Delta\mu_R \tag{6}$$

As for disubstituted benzenes of R–M–R′ type, the EIF model and the additivity hypothesis of the effects of the radicals R and R′ do not account satisfactorily for the observed values when R and R′ are of opposite nature (one acceptor, the other donor). In consequence an additional contribution β_{ct} due to internal donor–acceptor charge transfer has to be considered

$$\beta = \beta_{EIF} + \beta_{ct}. \tag{7}$$

Taking β_{ct} into account leads to a satisfactory agreement for benzene as well as for styrene, stilbene and phenyl butadiene derivatives. All the experimental results support the very simple idea that large values of β are associated with long chains of delocalised electrons strongly distorted by strong donor and acceptor radicals respectively in substitution at each extremity of the chain; the longer the chain, the larger β. Recent calculations

(Zyss 1979), based on a semi-empirical treatment of the Hartree–Fock equations extended to all valence electrons and using an INDO procedure, confirm the preponderance of Π contributions to β. They are in agreement with the EIF model for monosubstituted benzenes.

Note that the ultraviolet dispersion of the donor–acceptor charge transfer contribution to β has been examined by Levine and Bethea (1978), while Lalama and Garito (1979) developed an all-valence-electron-self-consistent-field linear-combination-of-atomic-orbitals molecular-orbital procedure for calculating the magnitude and sign of β for disubstituted dipolar aromatic molecules.

Nonlinear optical phenomena are of particular interest in detecting structural phase transitions as well as studying differences in phases of materials. For ferroelectric phase transitions the linear optical susceptibility usually changes by a few per cent while **d** may change by orders of magnitude thereby leading to very sensitive and precise detection techniques. It has also been shown that in ferroelectrics there is a linear relationship between the vector part of **d** and the spontaneous polarisation $\mathbf{P}_s$ (Jerphagnon 1970)

$$\mathbf{d}_v = K\mathbf{P}_s. \tag{8}$$

This relationship is not limited to the case of ferroelectrics. It also applies to the pyroelectric materials (i.e. having a polar crystallographic structure) for which it is difficult to make a good determination of $\mathbf{P}_s$. The building block of the wurtzite-type binary compounds (AlN, CdS, BeO, ZnO . . .) is a distorted tetrahedron which deviates from a regular one by a compression along the c axis leading to a reduced value of the ratio c/a and by a displacement u off the centre of the atom inside the tetrahedron. As a consequence of equation (8) the vector part $\mathbf{d}_v$ is a function of c/a and u

$$\mathbf{d}_v = f(c/a, u) \tag{9}$$

For studying phase transitions one can use the temperature dependence of the nonlinear optical susceptibility **d** to monitor the dependence of some structural or physical property and equation (8) can be useful in that respect. More important is the work of Bergman *et al* (1977 and references therein) who took advantage of their three-dimensional model to deduce information on the molecular mechanics of phase transitions.

Making equation (5) explicit, one obtains

$$\mathbf{d} = \sum_{U} \beta_U f_U(\phi) \tag{10}$$

where ϕ is a deformation or rotation angle for the building unit U. Assuming that the variation of the nonlinearity β_U is considerably less than the change in $f_U(\phi)$ with temperature, there is a direct relationship between the macroscopic nonlinearity and ϕ.

$$\mathbf{d} = kf_U(\phi) \tag{11}$$

The method has been successfully applied to various cases of phase transformations which include:

- Trigonal distortion of an octahedron: $LiNbO_3$, $LiTaO_3$
- Tetragonal distortion of an octahedron: $PbTiO_3$
- Rotation of a tetrahedron: GeO_4 in the ferroelectric transition of $Pb_5Ge_3O_{11}$, SiO_4 in the α–β phase transition of quartz.

4. Conclusion

Considerable progress has indeed been made lately in explaining the origin of the optical nonlinear susceptibilities in dielectric solids, and satisfactory models have been proposed for a large variety of materials. As clearly shown by the experiments already performed, nonlinear optics can give a new insight into the mechanisms of phase transitions and allow the study of ultrafast dynamical processes in solids. There is no doubt that the burgeonning field of nonlinear spectroscopy will have a considerable impact on dielectric physics by, for instance, providing us with powerful new tools for investigating the properties of dielectric solids.

References

Armstrong J A, Bloembergen N, Ducuing J and Pershan P S 1962 *Phys. Rev.* **127** 1918

Bergman J G and Crane C R 1974 *J. Chem. Phys.* **60** 2470

Bergman J G, Crane C R and Turner E H 1977 *J. Solid State Chem.* **21** 127

Bloembergen N 1965 *Nonlinear Optics* (London: Benjamin)

— 1977 *Proc. Int. School of Physics Enrico Fermi, Nonlinear Spectroscopy* (Amsterdam: North-Holland)

Chemla D S 1971 *Phys. Rev. Lett.* **26** 1441

Chemla D S, Oudar J L and Jerphagnon J 1979 *Nonlinear Behaviour of Molecules, Atoms and Ions in Electric, Magnetic or Electromagnetic Fields* (Amsterdam: Elsevier)

Davydov B L, Derkacheva L D, Dunina V V, Zhabotinskii M E, Zohin V F, Koreneva L G and Samokhina M A 1970 *JETP Lett.* **12** 16

Denbigh K G 1940 *Trans. Faraday Soc.* **36** 936

Faust W L and Henry C H 1966 *Phys. Rev. Lett.* **17** 1265

Flytzanis C 1975 *Quantum Electronics: a Treatise* vol 1 *Nonlinear Optics* ed. M Rabin and C L Tang (New York: Academic Press) p9

Flytzanis C and Ducuing J 1968 *Phys. Rev. Lett.* **26A** 315

— 1969 *Phys. Rev.* **178** 1218

Garret C G B and Robinson F N H 1966 *IEEE J. Quantum Electron.* **QE-2** 328

Jeggo C R and Boyd G D 1970 *J. Appl. Phys.* **41** 2741

Jerphagnon J 1970 *Phys. Rev.* **B2** 1091

— 1971 *IEEE J. Quantum Electron,* **QE-7** 42

Jerphagnon J, Chemla D S and Bonneville R 1978 *Adv. Phys.* **27** 609

Johnston W D Jr 1970 *Phys. Rev.* **B1** 3494

Kaminow I P 1967 in *Ferroelectricity* ed. E F Wheeler (Amsterdam: Elsevier)

Kleinman D A 1962 *Phys. Rev.* **126** 1977

Kurtz S K K *1966 Proc. Int. Conf. on Ferroelectrics* (Prague: Inst. Phys. Czech. Acad. Sci.) vol 1, p413

Kurtz S K K and Perry T T 1968 *J. Appl. Phys.* **39** 3798

Lalama S J and Garito A F 1979 *Phys. Rev.* **A 20** 1179

Lax B, Mavroides J and Edwards D 1962 *Phys. Rev. Lett.* 8 166

Levine B F 1969 *Phys. Rev. Lett.* **22** 787

— 1973 *Phys. Rev.* **75** 2591, 2600

Levine B F and Bethea C G 1978 *J. Chem. Phys.* **69** 5240

Miller R C 1964 *Appl. Phys. Lett.* **5** 17

Phillips J C 1969 *Covalent Bonding in Crystals, Molecules, and Polymers* (Chicago: University of Chicago)

— 1973 *Bonds and Bands in Semiconductors* (New York: Academic Press)

Rabin M and Tang C L 1975 *Quantum Electronics: a Treatise* vol 1 *Nonlinear Optics* (New York: Academic Press)

Robinson F N H 1967 *Bell Syst. Tech. J.* **46** 913

Scholl F and Tang C L 1973 *Phys. Rev.* **B8** 4607
Tang C L 1973 *IEEE J. Quantum Electron.* **QE-9** 755
Wemple S H and Di Domenico M 1968 *Appl. Phys. Lett.* **12** 352
— 1972 *Applied Solid State Science: Advances in Materials and Device Research* vol 3, ed. R Wolfe (New York: Academic Press)
Zernike F and Midwinter J E 1973 *Applied Nonlinear Optics* (New York: Wiley)
Zyss J 1979 *J. Chem. Phys.* **70** 3333

Inst. Phys. Conf. Ser. No. 58
Invited paper presented at Physics of Dielectric Solids, 8–11 September 1980, Canterbury

Electronic transport in amorphous dielectric films

J Mort
Xerox Corporation, Webster Research Center, Webster, New York 14580, USA

Abstract. Over the last few years considerable progress has been made in the study of electronic transport in a wide range of amorphous dielectric materials. These embrace chalcogenides such as a-Se and a-As_2Se_3, tetrahedrally bonded materials such as hydrogenated a-Si and organic systems such as molecularly doped polymers. A comparative survey is given of selected transport phenomena in these three classes of amorphous solids but with particular emphasis on the last. The unifying theme is the role of localised states and how they influence and control photoelectronic processes such as electronic transport and its sensitivity to doping. Despite similarities, however, features are observed which are unique to each class and reflect their specific chemical nature.

1. Introduction

Amorphous dielectric thin films are of both scientific and technological interest, and in fact the technological interest and successful exploitation spawned the scientific investigations. These scientific studies have been concerned with the question as to the effects of disorder on the solid state properties. Almost immediately it became clear that disorder had the biggest impact on the electrical and photoelectronic properties of solids. The general reasons for this were intuitively understood since disorder introduces intrinsic localised tail states at the band edges. Electrical properties which involve spatial displacement of charge are therefore extremely sensitive to the presence of such states. The picture however is complicated by the presence of deeper lying gap states related to defects associated with dissatisfied bonds or ubiquitous impurities. Initially therefore experimentalists and theorists were faced with complex and confusing observations. From the experimental point of view the situation was further complicated by the limited experimental techniques that could be applied to amorphous insulating films. Thus one of the most important experimental tools in understanding the electrical properties of crystalline semiconductors, the Hall effect, has been applied with only limited success to amorphous solids. Fortunately another method also first applied to crystalline solids, *viz* the time-of-flight technique, has proven to be the most important and direct way of studying electronic transport in amorphous dielectric films.

This paper will deal with the particular issue of the time-of-flight technique and the results that have been obtained in three general groups of amorphous materials, the chalcogenides, the tetrahedrally bonded solids and molecular systems. These together make up the majority of dielectric thin film materials. Only aspects of electronic transport will be considered and in this paper no discussion will be given of related phenomena such as charge photogeneration.

0305-2346/81/0058-0082 $01.50

2. Time-of-flight techniques

The basic ideas of the time-of-flight technique are illustrated schematically in figure 1. A sample of the material under study, of thickness L and capacitance C, is sandwiched between electrodes which should be blocking. A short pulse of strongly absorbed radiation passes through a semi-transparent electrode and generates a thin sheet of carrier pairs. If the dielectric relaxation time is longer than any subsequent time of observation then the carrier pairs can be separated and the independent drift of either carrier studied by choice of the appropriate polarity of drift field. Providing the time to drift across the sample thickness L is much less than the deep trapping lifetime τ, then the transit time t_T can be found by measuring the duration of the induced current or charge displacement. If the number of carriers drawn out of the excitation region is Q and $Q \ll CV$, then in the absence of any deep trapping, a constant current equal to Q/t_T flows during the transit. Thus the duration of the transit allows a determination of the drift mobility $\mu_d = L^2/t_T V$ while the integral of the transit pulse allows a determination of Q. Together with a knowledge of the number of absorbed photons this allows a measurement of the photogeneration efficiency if the absence of carrier losses within the excitation region can be established.

The transit can be monitored in either the current or voltage mode depending on the relative magnitudes of the transit time t_T and the external circuit RC time constant. If $RC \ll t_T$, a rectangular current pulse is observed as in figure 1, whereas if $RC \gg t_T$, its integral, a ramp function, is the observed response. A choice between the two modes depends on whether detailed pulse shape information is sought in which case the current mode is more revealing, or signal sensitivity is critical, in which case the voltage mode is inherently more sensitive. In addition to pulse excitation it is also possible to employ step-function excitation. In this case the response in the *current mode* is a ramp function

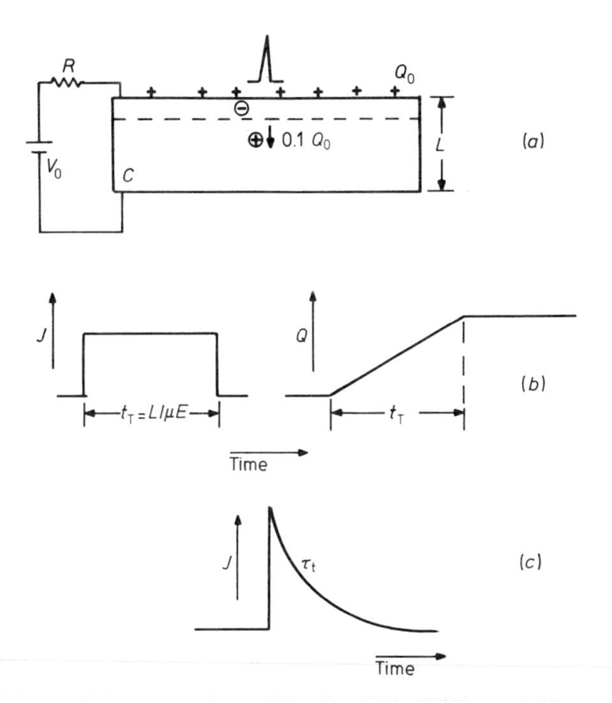

Figure 1. Schematic illustration of idealised time-of-flight experiment: (*a*) sample of thickness L and capacitance C is charged to surface charge density $Q_0 = CV_0$; Light flash induces small perturbation $<0.1Q_0$ by photogeneration; (*b*) rectangular current pulse or ramp-shaped charge displacement pulse corresponding to $RC \ll t_T$ or $RC \gg t_T$ respectively; and (*c*) exponential decay due to deep trapping.

and the steady current at low excitation level is a direct measure of the rate of carrier supply out of the excitation region, the so-called emission-limited current. The step-function excitation can be produced either by step-function illumination or by using steady illumination to produce a reservoir of carriers which can be injected by a step-function voltage pulse.

The transient waveforms discussed so far are only observed under idealised conditions and are rarely encountered in practice although a-Se at room temperature is an excellent example of the case where these conditions are closely met. The most general reason for departure from the idealised case is due to carrier trapping. It is convenient in this respect to distinguish between shallow and deep trapping processes. The distinction between these two cases is somewhat arbitrary since it rests on the probability of re-emission during the time of observation. The release time τ_r is exponentially dependent on the trap depth and during this release time when the carrier resides in the trap it is immobilised. An equally critical time in this process is the trapping time τ_t. If $\tau_t \ll t_T$ and $\tau_t \ll \tau_r \ll t_T$ then most of the carriers will undergo many trapping events during a transit. The effective transit time will therefore be increased and the drift mobility is a multiple-trap-controlled process μ_d where

$$\mu_d = \mu_0 [1 + N_t/N_c \exp(\Delta E/kT)]^{-1} \tag{1}$$

where μ_0 is the carrier mobility without trapping, the microscopic mobility, N_t is the density of traps, N_c is the effective density of states at the band or mobility edge and ΔE is the energy of the trap with respect to the band or mobility edge. This information concerning N_t and ΔE can in principle be determined by a study of the temperature dependence of μ_d.

Deep trapping occurs when $\tau_r \gg t_T$ and is an important parameter in two respects. First, it is critical in determining many photoconductor applications since, for example, in electrophotography it determines the distance a carrier can move in a given field ($\mu\tau_t E$) and therefore the degree of photodischarge (Mort and Chen 1975). Second, it determines whether the time-of-flight technique can be successfully applied to a given material. Thus if $\tau \ll t_p$ where t_p is the resolution time of the apparatus then it will be impossible by any variation of sample thickness, applied voltage or temperature to observe a transit pulse. If a well defined single deep trapping lifetime exists then it manifests itself as an exponentially decaying current pulse with a time constant equal to τ_t. It must be stressed that the terminology of shallow and deep traps does not necessarily apply to a particular trap under all conditions. Thus a trap may function as a shallow or deep trap depending on temperature.

These trapping effects, which can result in departures from the idealised features, are general in nature and apply to both crystalline and non-crystalline material. In the case of non-crystalline or amorphous materials another, rather more fundamental, reason for non-idealised behaviour exists. A rectangular current pulse can only be expected if a constant number of carriers move with a single-valued constant velocity. This means that a unique mobility or mean velocity per unit field can be defined. Such gaussian transport is associated with one dominant event time. In amorphous materials the transport process may (or may not, as is the case for a-Se at room temperature) involve a wide distribution of individual event times. Such a stochastic transport process has been mathematically described by Scher and Montroll (1975). Although this theory is of general applicability and is not constrained to transport that involves hopping, this transport process will serve as an illustrative example. As a result of statistical variations in hopping rates, carriers that initially start as a well defined sheet ultimately spread out throughout the sample. Thus, for example, some carriers can immediately hop out of the excitation region while many carriers remain within this region. After each hop a carrier can encounter a site from which the probability for the next hop is extremely small. With the passage of time an increasing number of carriers undergo this experience. As a result, even before any carriers transit and therefore leave the sample, the current progressively decreases even though the total number of carriers in motion remains constant. Eventually the carriers which have the fastest transit time reach the back electrode and carriers then begin to be lost to the system. The time at which this occurs is operationally defined as the transit time t_T. For a sufficiently broad distribution of hopping probabilities the maximum of the charge distribution remains close to the generation region even for times greater than t_T. The resultant dispersion of the carrier sheet and the mean displacement of the charge distribution from the generation region grow with time in the same way and their ratio remains constant. This predicts the so-called universality of the transient pulse shape in which the distribution of charge within the transit pulse is independent of the time domain in which the transit takes place. Such universality does not necessarily occur if the transit time is changed by altering the temperature.

Scher and Montroll (1975) have shown that the probability for a carrier to jump to the next site at time t after having arrived at $t=0$ is a slowly decaying function which can be approximated by

$$\psi(t) \simeq t^{-(1+\alpha)} \tag{2}$$

where α is a disorder parameter. It was shown that the transient current decays as

$$i \sim \begin{cases} t^{-(1-\alpha)}, & t < t_T \\ t^{-(1+\alpha)}, & t > t_T \end{cases} \tag{3}$$

The increase in the power exponent of the decay at $t \simeq t_T$ occurs when the leading edge of the carrier sheet encounters the absorbing substrate electrode. In logarithmic scales, the current trace as indicated in figure 2 should appear as two straight lines intersecting at t_T with an initial slope of $-(1-\alpha)$ and a final slope $-(1+\alpha)$. It should be stressed that this transit time is not a unique property of the carrier sheet and represents only the transit time of the fastest carriers, which may be only ~20% of the total. Thus no mean velocity per unit field or mobility can be defined. The transient pulse in fact reflects the velocity distribution itself. Operationally defined mobilities are predicted on the basic of this model to be field dependent, even in the absence of a field-dependent transition probability, and thickness dependent. Nevertheless the time t_T is related to the transport mechanism and is an operationally useful fiduciary. In many cases t_T can be discerned in logarithmic plots of the transit pulse when in linear scales the transient pulse is completely featureless (Pfister and Scher 1978).

3. Experimental results

3.1. Molecularly doped polymers

Many polymers are amorphous in character and exhibit charge transport features that are in many respects remarkably similar to inorganic amorphous solids. A fundamental

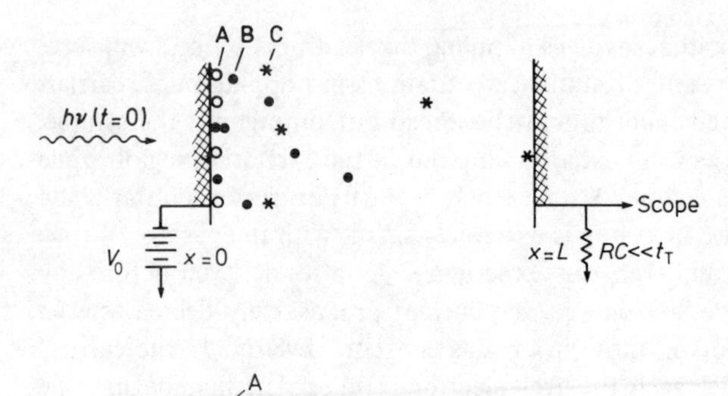

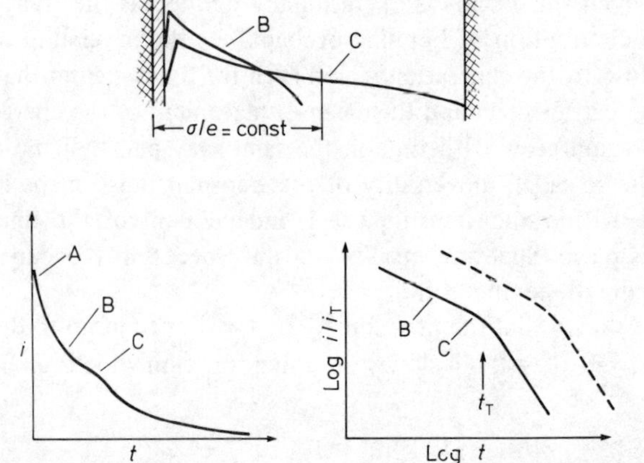

Figure 2. Schematic representation of carrier propagation under ideal non-gaussian conditions. Top: position of representative carriers in the sample bulk at t=0(○), t<t_T(●), and t~t_T(*). Middle: charge distribution in sample bulk at t=0, t<t_T, and t~t_T. Bottom: current pulse in external circuit in linear units (left) and logarithmic units (right). Broken curve represents transient current for lower applied bias field, i.e. longer transit time (after Pfister and Scher 1978).

difference however is that, in the case of pendant group polymers or molecularly doped polymers in which chromophores are either attached to the polymer chain or dispersed in a polymer matrix respectively, the localised states are molecular in nature (Mort 1980a).

The electronic absorption spectra of such polymer systems are similar to those of the polymers in solution or of effectively isolated chromophores in the gas phase (Ritsko and Bigelow 1978). This is evidence that the weak interaction between the pendant or doped molecules leads to small exchange energies and bandwidths much smaller than the energy fluctuations associated with the disorder. In general, such electronic states will be localised. The pendant molecules have a much lower ionisation potential than the saturated polymer backbone and are expected and observed to dominate the low-energy transitions and to play an integral role in their electrical properties. The energy required to separate a negative and a positive charge for an isolated molecule (as in gas-phase ionisation) is $I_g - A_g$, where I_g is the gas-phase ionisation energy and A_g is the energy gained by binding the ionised electron to a neighbouring homologue molecule. In a solid the energy ΔE to produce these ion states is diminished by the existence of the molecules within a polarisable medium; thus $\Delta E = I_g - A_g - 2P$, where P is the polarisation energy of the 'free electron' and, to a good approximation, is the same for electrons and holes. In this sense, one can speak of conduction states for electrons and holes separated by a band gap, ΔE (Guttman and Lyons 1967).

Extensive time-of-flight and xerographic discharge measurements have been carried out on polymers based on carbazole such as poly(N-vinycarbazole) (PVK) (Mort 1972, Pfister and Griffiths 1978), their charge transfer complexes, for instance, with trinitrofluorenone

(TNF) (Gill 1973), and molecularly doped polymers (Mort and Pfister 1979). The latter are solid molecular solutions of a polymeric matrix such as polycarbonate (Lexan) and a dopant molecule such as triphenylamine (TPA) (Pfister 1977), or N-isopropyl carbazole (NIPC) (Mort *et al* 1976). In all these systems the same basic transport properties prevail. The charge carriers propagate through the solid by a hopping mechanism which proceeds among the chromophores in the intrinsically photoconducting polymeric solids and among the dopant molecules in the molecularly doped polymers. Pioneering work in intrinsically conducting polymers was performed by Gill (1973) who investigated in detail the transport properties of PVK and PVK: TNF. These studies led to the concept of molecular doping. By introducing carbazole side groups in the form of N-isopropylcarbazole into polycarbonate host polymers, Mort *et al* (1976) demonstrated that the vinyl backbone in the PVK polymer does not contribute to charge transport but merely provides the mechanical stability of the films. By comparison with the intrinsically conducting polymers, the molecularly doped polymers are much more flexible in material design since the concentration and kind of dopant molecule can be selected, which allows one to optimise certain transport properties. To some extent the mechanical properties of the host polymer can also be chosen. This flexibility proves to be invaluable in unravelling the details of charge transport (and generation) in these systems. In what follows, molecularly doped polymers are used to illustrate features involved in charge transport in organic disordered solids and which are fundamental to the disordered state in general.

The basic properties of transport in organic disordered solids are illustrated using as the model system TPA molecularly dispersed in bisphenol-a-polycarbonates (Lexan). Transport studies were performed using the time-of-flight technique (Pfister 1977) and the technique of space-charge-limited xerographic discharge (Borsenberger *et al* 1978). In the former measurements the free carriers are generated by a 5 ns laser pulse of wavelength 337.1 nm which is exclusively and strongly absorbed by the dopant molecule or in a thin a-Se sensitising layer.

No complex formation is detected in the solid film, i.e. the total absorption equals the sum of the absorption of polymer host and dopant molecule. Therefore, the latter can be excited selectively. The films, 10–15 μm thick, are cast from solution onto aluminium substrates using a draw-down coating apparatus and the solvent, usually 1,2-dichloroethane, driven off by drying in vacuum. The TPA concentration is reported as the weight ratio X of triphenylamine to polycarbonate as introduced into the solution.

Only hole transport can be observed in TPA/Lexan. Figure 3 shows the hole drift mobility as a function of temperature in the form of an Arrhenius plot, $\log \mu$ versus $1/T$. The data were recorded for an applied field of 70 V μm^{-1} and various TPA concentrations.

Three main conclusions can be drawn from figure 3. The hole mobility is thermally activated with an activation energy Δ that increases with decreasing TPA concentration. Compare, for instance, $X = 0.5, \Delta = 0.32$ eV and $X = 0.1, \Delta = 0.49$ eV. For constant temperature the mobility rapidly drops as the TPA concentration is reduced. Thus, at $T = 296$ K the mobility decreases from $\sim 10^{-5}$ $cm^2 V^{-1} s^{-1}$ for $X = 0.5$ to $\sim 3 \times 10^{-9}$ $cm^2 V^{-1} s^{-1}$ for $X = 0.1$.

The strong dependence of the hole mobility on TPA concentration indicates that charge transport occurs by a hopping process. In this case the mobility is proportional to the overlap integral of the wavefunction that localised the hole carrier on neighbouring dopant molecules. For such a process one expects the approximate relation $\mu \propto \rho^2 \exp(-2\alpha\rho) \exp(-\Delta/kT)$ where ρ is the average separation of the TPA molecules which are

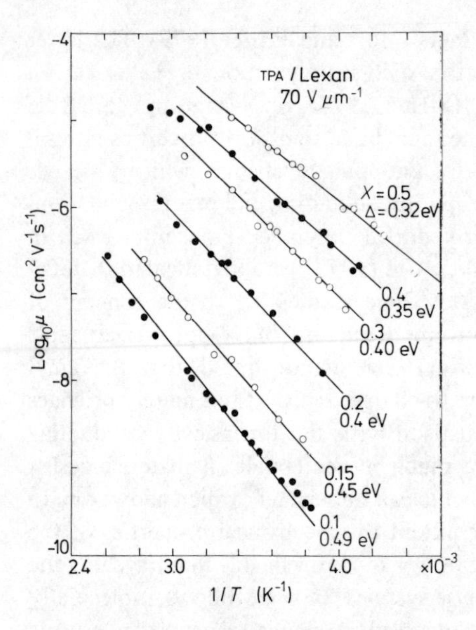

Figure 3. Hole drift mobility in TPA/Lexan as a function of temperature. The data were taken at 70 V μm^{-1} applied field for various TPA concentrations, X (after Pfister 1977).

assumed to be uniformly dispersed and α^{-1} is a parameter that describes the exponential decay of the wavefunction outside the molecule. To a first approximation, one may estimate ρ from the TPA concentration N_{TPA} using $\rho \sim (N_{\mathrm{TPA}})^{-1/3}$.

Figure 4 shows the exponential dependence of the hole mobility on the separation between dopant molecules for a number of organic systems including the PVK : TNF charge transfer complex. In the latter, holes hop among the carbazole side groups. Complexing with TNF removes carbazole units from the hopping channel, and consequently, the average hopping distance increases Gill (1973). From the slope of the lines in figure 4 one obtains the 'localisation' parameter α^{-1} that characterises the decay of the wavefunction

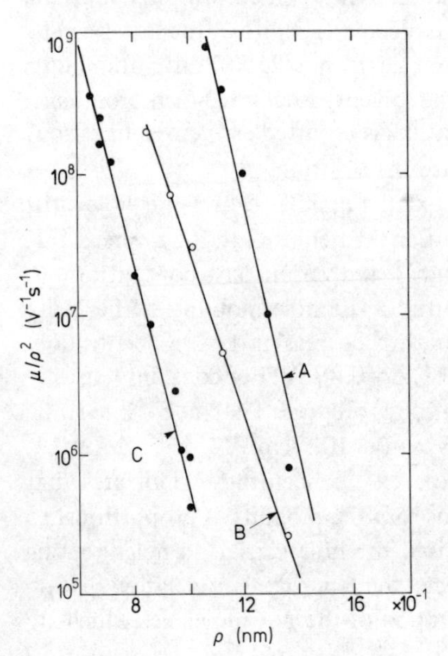

Figure 4. Concentration dependence of hole drift mobility in the molecularly doped polymers (A) TPA/ Lexan (α^{-1} = 0.1 nm) and (B) NIPC/Lexan (α^{-1} = 0.15 nm) and (C) the charge transfer complex PVK : TNF (α^{-1} = 0.11 nm) after Mort *et al* (1976) and Gill (1973), at 50 V μm^{-1} and 296 K.

outside the molecule. Using the conventional definition for the localisation radius for hopping among point sites, one obtains from figure 4 the values $\alpha^{-1} \sim 0.11$ nm, 0.15 nm, and 0.10 nm for PVK:TNF, NIPC/Lexan, and TPA/Lexan, respectively. This is an extremely small value as compared to the localisation of charge on donor sites in a-Si as discussed later.

The drift mobility and its activation energy are field dependent. Figure 5 shows log μ against log E for TPA/Lexan doped at the level $X = 0.5$. In the field range 20–80 V $\mu\mathrm{m}^{-1}$, one finds that the activation energy decreases with applied field approximately as $\Delta(E) \sim \Delta_0 - \beta E$. β increases slightly with increasing hopping distance ρ but its value for fixed concentration does not significantly depend upon the kind of dopant molecule. This is of interest since the field dependence of the drift mobility at room temperature has been observed to vary from $\mu \sim$ const to $\mu \propto E^3$ for various doped polymer systems. On the other hand, the zero field intercept Δ_0 can significantly change with concentration and polymer system. A typical range is $\Delta_0 \sim 0.3$ to 0.7 eV. This range of Δ_0 would amount to a variation of the drift mobility by a factor of $\sim 10^7$ as is estimated for $\exp(-\Delta/kT)$ for room temperature.

For all systems reported so far, the field dependence of the mobility is much weaker than would be expected on the basis of the Arrhenius temperature term $\exp(-\Delta(E)kT)$ calculated for the measured value $\Delta(E)$. This deviation is observed at all temperatures. Hence the strong exponential field dependence has to be offset by a field dependence which to a first approximation is temperature independent. This can be done by introducing in the Arrhenius term an effective temperature, T_{eff}, rather than the laboratory temperature T, where $T_{\mathrm{eff}} > T$. An extensive study of hole transport in TPA/Lexan (Pfister 1977) and the charge transfer complex PVK:TNF (Gill 1973) indicates the phenomenological relationship $1/T_{\mathrm{eff}} = 1/T - 1/T_0$ which applies for $T < T_0$. T_0 is a temperature operationally defined, which appears characteristic of the hole transport system (Pfister 1977). Hence the smaller the difference between T and T_0, the larger the effective temperature T_{eff} and

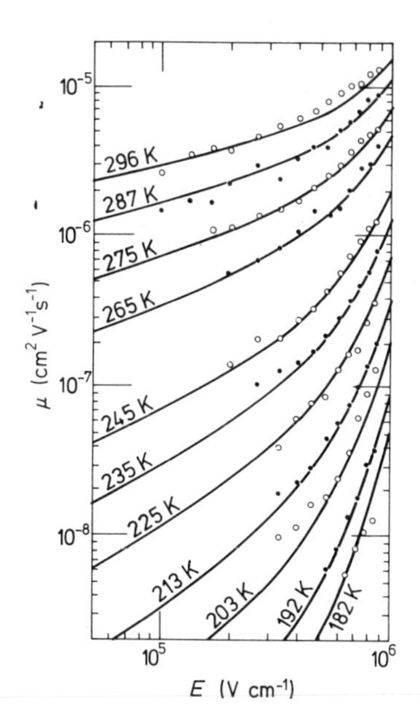

Figure 5. Field dependence of hole drift mobility in TPA/Lexan at various temperatures (after Pfister 1977). 0.5 TPA, $\alpha_0 = 0.26$, $T_0 = 455$ K, $\Delta_0 = 0.42$ eV, $\rho = 1.07$ nm.

the weaker the field dependence $\mu(E)$ that is calculated from $\Delta(E)$. For instance, $\mu(E)$ varies more strongly for PVK than for TPA/Lexan. The analysis of the temperature dependence of the hole mobility yields $T_0{\sim}550$ K for PVK and $T_0{\sim}455$ K for a TPA/Lexan sample of 0.5 weight ratio which is in qualitative agreement with the relative strength of the field dependence $\mu(E)$ for the two materials. These arguments suggest that molecules that give rise to weakly field-dependent mobilities are characterised by low T_0 (or high T_{eff}). Hence T_0 is an important parameter in the study of hole transport in molecularly doped polymer systems.

The strong dependence of the hole drift mobility upon intersite distance ρ suggests that the carriers propagate by a hopping motion. This mechanism can be visualised as either an oxidation–reduction or a donor–acceptor process. Under the influence of the applied electric field, neutral molecules will repetitively transfer electrons to their neighbouring positively charged radical cations (figure 6). The net result of this process is the motion of a positive charge across the bulk of the sample film. Note that this is strictly an electronic and not an ionic transport process since no mass displacement is involved. Hence, for hole transport to occur, one expects that the dopant molecule is donor-like in its neutral state. On the other hand, for electron transport, where the electron hops from the radical anions to their neighbouring neutral molecules, the dopant molecules are acceptor-like. Indeed, for the donor-like molecules NIPC and TPA dispersed in polycarbonate, only hole transport is observed (Mort *et al* 1976, Pfister 1977, Borsenberger *et al* 1978), while for the acceptor molecule TNF dispersed in polyester, transport only occurs for electrons (Gill 1973). Thus electronic transport in the disordered organic state is typically unipolar and the sign of the mobile carriers is determined by the molecular properties of the dopant molecule. The situation is completely different in molecular crystals which typically are ambipolar with both carriers having similar mobilities (~1 cm^2 V^{-1} s^{-1}).

In the ideal hopping process, one thinks of a charge hopping among point-like localised sites whose geometrical dimensions are much less than their average separation. These conditions are met in the doped polymers or intrinsically semiconducting polymers. The localised sites are dopant molecules or side groups attached to a polymer backbone, and the spatial dimensions of these groups are typically in the range 0.15–1.0 nm. This is to be compared with the average separation of 1–2 nm calculated from the density of the molecules as they are introduced into the films. Due to the proximity and nonspherical shape of the transport active molecules, one expects that the integral of the wavefunctions of a carrier localised on neighbouring molecules, the overlap term, is very sensitive to the relative orientation of adjacent molecules. The overlap term can vary by many orders of

Hole transport

e

m m⊕ m m

Electron transport

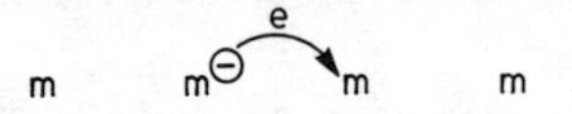

Figure 6. Schematic representation of transport process in molecularly doped polymers.

magnitude as a function of the relative orientation of two molecules for a fixed distance of the respective centres of mass. It is to be expected then that the thermal motions of molecules modify the overlap term. leading to stronger localisation at lower temperatures.

The observation of a temperature-dependent overlap $\alpha(T)$ suggests an interesting microscopic transport step. Thus, if a carrier arrives at a molecule whose overlap with neighbouring molecules is extremely poor due to mutual misalignment (such a situation might be called a 'conformational trap' (Slowik 1976)) the hopping time to the next molecule might be longer than the time it takes to optimise the overlap by relative molecular motion of adjacent molecules. Hence the carrier would await such a coincidence and then rapidly complete the hop. Since the rate-limiting step now involves molecular motion, one expects that its activation energy contains a term reflecting a barrier to molecular motion. Indeed, the typically large transport activation energies observed in disordered organic solids suggest that such contributions add to the energies of electronic polarisation (polaron binding energy) and disorder.

A further contribution to the activation energy can also result from trapping. Since the transport in disordered organic polymeric solids is typically unipolar, one can predict that only those impurities with an ionisation potential less than that of the dopant molecule will be hole trapping sites. An experimental verification of this prediction has been reported for the hole transport system NIPC/Lexan into which, for fixed NIPC concentration, various amounts of TPA were introduced (Pfister *et al* 1976). The pertinent results are shown in figure 7.

Two sets of samples were measured. In one set the NIPC concentration was zero and the TPA concentration was varied as described before. The strong concentration dependence of the hole velocity shown by the solid circles in figure 7 reflects the hopping transport previously discussed. In the other series of samples, Lexan was doped with NIPC molecules (with $n_{\mathrm{NIPC}} = 1 \times 10^{21}\ \mathrm{cm}^{-3}$) and the TPA concentration was varied. n_{TPA} was kept below $\sim 3 \times 10^{20}\ \mathrm{cm}^{-3}$ to ensure that the average intersite distance $\rho_{\mathrm{NIPC}} \sim (n_{\mathrm{NIPC}})^{-1/3}$ among the NIPC molecules was not changed by the TPA molecules. The drift velocity results at $50\ \mathrm{V}\ \mu\mathrm{m}^{-1}$ are shown in figure 7 as the open circles. For $n_{\mathrm{TPA}} = 0$ (arrow),

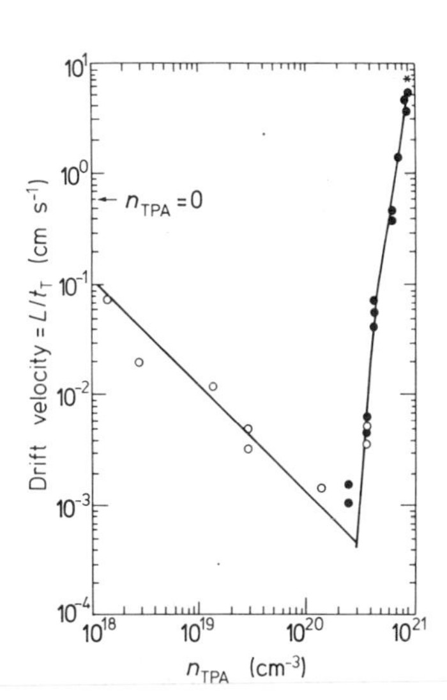

Figure 7. TPA concentration dependence of hole velocity in Lexan (•) and NIPC/Lexan where the NIPC concentration was fixed at $1 \times 10^{21}\ \mathrm{cm}^{-3}$ (○) and $2 \times 10^{20}\ \mathrm{cm}^{-3}$ (*) (after Mort and Pfister 1979).

transport occurs by hopping among NIPC molecules. Note that at the same concentration of $\sim 1 \times 10^{21}$ cm^{-3}, transport via TPA molecules exceeds that via NIPC molecules by more than one order of magnitude.

The addition of TPA reduces the drift velocity from the value at $n_{TPA} = 0$ in a manner which, for $n_{TPA} \sim 10^{18}$ to 10^{20} cm^{-3}, is approximately proportional to n_{TPA}^{-1}. With further increasing TPA concentration, the velocity goes through a minimum at $\sim 3 \times 10^{20}$ cm^{-3}, the value obtained with films of the first sample series which contained no NIPC. Samples with $n_{TPA} > 4 \times 10^{20}$ cm^{-3} cannot be prepared without changing the average intersite distance among NIPC molecules. Also, for total concentrations in excess of 2×10^{21} cm^{-3}, crystallisation effects become apparent.

From these concentration studies alone it is possible to explain the general features of the observations in a relatively simple way. For $n_{TPA} = 0$, the hole transport occurs via hopping among the NIPC molecules present in a fixed concentration. As TPA molecules are introduced at low concentrations, carriers occasionally become localised on a TPA molecule, which, because its ionisation potential is lower than NIPC, acts as a trap for holes. Since the overlap between TPA molecules is so small at these concentrations, further drift of the charge localised on TPA must await thermal excitation back to an NIPC molecule. The data points for $n_{TPA} < 2 \times 10^{20}$ cm^{-3} pertain to this mechanism. At sufficiently high TPA concentrations the overlap among the TPA molecules becomes large enough that TPA–TPA hopping begins to compete with the hopping among NIPC–NIPC and TPA–NIPC pairs observed at low TPA loading. This process causes the drift velocity to rise for $n_{TPA} > 2 \times 10^{20}$ cm^{-3}, and it appears from the data shown in figure 7 that $n_{TPA} \sim 10^{20}$ cm^{-3}, hopping among TPA completely dominates the charge transport.

In low concentrations TPA inhibits hole transport through NIPC because its ionisation potential is lower. It follows then that, in the converse case, charge transport through TPA should not be influenced by NIPC as long as the intersite distance $(n_{TPA})^{-1/3}$ remains constant. This is experimentally confirmed in figure 7 by the coincidence of the velocities measured for $n_{TPA} \sim 4 \times 10^{20}$ cm^{-3} for both sample series and by the point identified by an asterisk which pertains to a sample with the loadings $n_{NIPC} \sim 2 \times 10^{20}$ cm^{-3} and $n_{TPA} \sim 8.5 \times 10^{20}$ cm^{-3}.

These experiments clearly demonstrate the strength of the molecularly doped systems for the interpretation of transport data on a microscopic level. In inorganic disordered solids the identification of transport mechanisms has often to rely upon numerical arguments. For the organic systems the ability to control the densities and species of the localised state produces unambiguous evidence of the underlying transport process.

3.2. Chalcogenides

The most widely studied chalcogenides are a-Se and a-As_2Se_3 which are also both of technological significance in electrophotography. Despite their similarity they exhibit quite different transport properties which cover the range of possible phenomena expected.

The idealised rectangular-shaped current signal is closely approached by transients of both carriers in a-Se at room temperature (Spear 1957, 1960). Figure 8 shows a typical hole transit signal where the transit time t_T is easily identified. The observed dispersion on the falling edge of the pulse increases approximately at $t_T^{1/2}$ consistent with a transport process governed by gaussian statistics. The room temperature mobility of holes is ~ 0.14 cm^2 V^{-1} s^{-1} and it has an activation energy for temperatures below 250 K of

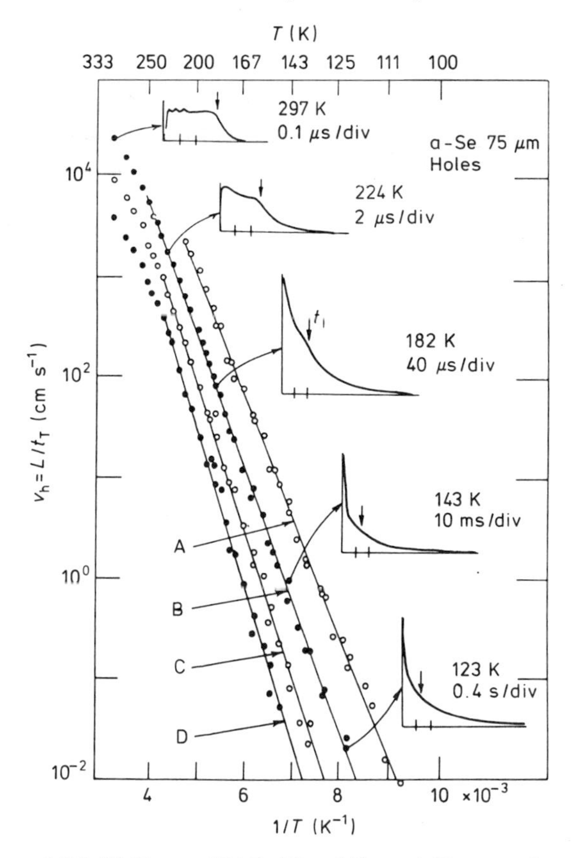

Figure 8. Activation energy plot of hole velocity in a-Se for various fields with representative transit pulse traces shown. The arrows mark the transit times determined from log–log plot analyses (after Pfister 1976). A, 20 V μm^{-1}; B, 10 V μm^{-1}; C, 5 V μm^{-1}; D, 2.5 V μm^{-1}.

~0.25 eV (Spear 1960). The drift mobility remains well defined with decreasing temperature down to 170 K below which the dispersion increases rapidly until finally transit times can no longer be discerned from currents displayed in linear scales (Pfister 1976). One must then resort to log–log representations to observe t_T. Universality is observed with respect to field but not temperature. One of the most intriguing aspects of these observations is the fact that despite the changes in current pulse shapes no change in activation energy is observed. With the onset of the dispersive transits the effective drift mobilities become both thickness- and field-dependent (Pfister 1976). The lack of change in activation energy suggests that the transport process remains unchanged but no unambiguous choice between hopping or multiple-trapping involving extended states is possible. Electron drift mobilities are ~5–8 × 10^{-3} cm^2 V^{-1} s^{-1} at room temperature and, over the temperature range for which data exists, exhibit very little dispersion. In both cases a specific choice between two competing transport processes, *viz* trap-controlled extended state or trap-controlled hopping, is not possible. Studies of doping effects on a-Se indicate that the effects of incorporating halogens or arsenic are in general to leave the drift mobility unaffected, but significantly change the deep trapping lifetime. Measurements of lifetimes indicate values ~<50 μs for electrons and holes in undoped a-Se (Tabak and Hillegas 1971).

In a-As_2Se_3 only hole transport has ever been observed and deep trapping lifetimes of the order of seconds have been reported (Scharfe 1970). In the case of a-As_2Se_3 the transport is typically very dispersive. Significant effects, due to doping, on the hole drift mobility are observed (see figure 9) with both increases and decreases depending on the nature of the dopant. The changes in drift mobility for some dopants are accompanied by corresponding changes in the dark conductivity (Pfister and Morgan 1980). For dopant levels below the alloy range, no evidence has yet been found for any significant shift in the Fermi level introduced by doping. This is somewhat surprising because other evidence, particularly from xerographic studies, suggests that the gap state densities in chalcogenides

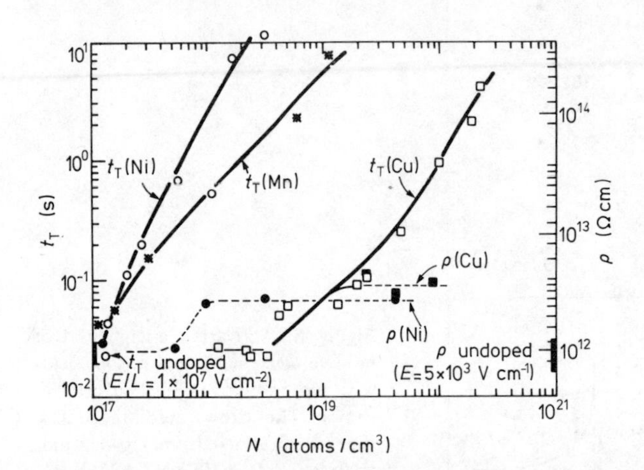

Figure 9. Log–log plot of transit time (left ordinate) and resistivity ρ (right ordinate) against concentration N of Ni-, Mn- and Cu-doped a-As_2Se_3 (after Pfister and Morgan 1980). ○, ●, Ni; *, Mn; □, ■, Cu.

are relatively low and certainly comparable to a-Si:H. The lack of a field effect is also consistent with an inability to move the Fermi level. This insensitivity of the position of the Fermi level in chalcogenides to doping is consistent with the defect models of Mott *et al* (1975) and Kastner *et al* (1978). Such models are based on valence alternation pair defects, D^+ and D^-, which are positively and negatively charged defects whose creation is governed by the law of mass action. Such defects produced under thermal equilibrium effectively pin the Fermi level. Recent studies confirm that the densities of defects can be affected by thermal treatment (Abkowitz and Enck 1980). Thus the photoelectronic properties of chalcogenides immediately after preparation are found to be dependent on thermal history. Characteristic equilibrium behaviour, independent of preparatory conditions, is achieved after annealing. This is consistent with Spear's observation (Spear 1974) that the most remarkable feature of the chalcogenides is their reproducibility with regard to drift mobility values, a feature probably unmatched by even the best crystalline semiconductors. This appears to be related to the ability of chalcogenide glasses to efficiently self-anneal defects.

3.3. Hydrogenated amorphous silicon

Hydrogenated amorphous silicon a-Si:H is receiving considerable attention because of its possible technological applications as an amorphous extrinsic semiconductor. Potential applications include electrophotography, photovoltaic cells and thin film electronic devices (Mort 1980b). In contrast to unhydrogenated a-Si, prepared for example by thermal evaporation, a-Si:H has a sufficiently high resistivity to permit the application of the time-of-flight technique (Spear and LeComber 1972).

The initial measurements of electron drift mobility in a-Si:H prepared by the glow discharge of silane are shown in figure 10. This data was interpreted in terms of trap-controlled extended state transport above about 250 K. Below this temperature a transition

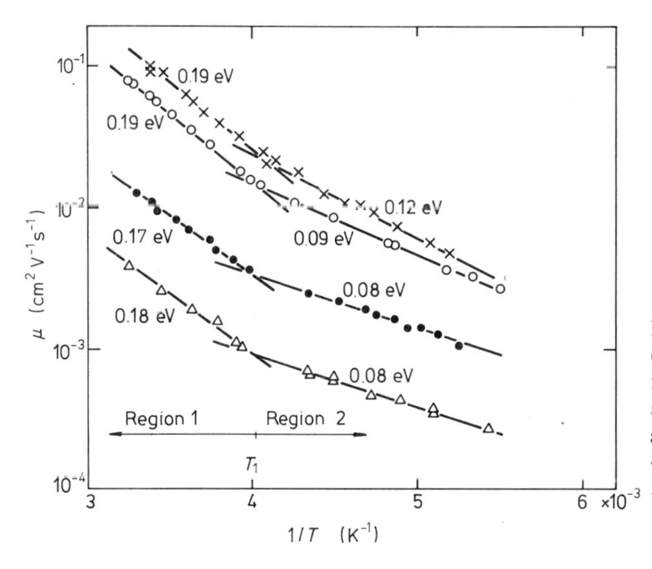

Figure 10. Temperature dependence of the electron drift mobility for four a-Si specimens prepared at the different substrate temperatures T_d; ×, 440 K; ○, 500 K; •, 460 K; △, 440 K; (after Spear and LeComber 1972).

to hopping conduction through tail states occurs (Spear and LeComber 1972). It can be seen that room temperature mobilities are ~0.1 cm^2 V^{-1} s^{-1} and recently values ~0.8 cm^2 V^{-1} s^{-1} have been reported by Tiedje *et al* (1980). Depending on sample preparation conditions the transient pulses can be either dispersive or nondispersive. A particularly interesting application of the time-of-flight technique has proved possible in a-Si:H because of the ability to extensively dope with donors or acceptors. Thus interdonor hopping transport has been directly observed in material compensated with boron acceptors which was necessary to maintain the high resistivity and sufficiently long dielectric relaxation time necessary to do the time-of-flight studies (Allan *et al* 1977). By control of the donor densities the average distance between donor sites, R_D ranged from 9.0 to 3.9 nm. Figure 11 shows a semi-logarithmic plot of μ_d/R_d^2 versus R_D. From the slope of this line an estimate of the spatial decay of the wavefunction away from the donor site of ~2.2 nm can be made. This is close to the value predicted for the hydrogenic model of donor states in crystals but is radically different from that found in disordered molecular systems as discussed earlier.

The drift mobility of holes in a-Si:H is much lower, being ~10^{-4}–10^{-5} cm^2 V^{-1} s^{-1} at room temperature, strongly thermally activated ~0.4–0.5 eV and is dispersive (Moore 1977, Mort *et al* 1980b). The mechanism of the transport process is unclear and either trap-controlled extended state transport or trap-controlled hopping are both possible. Recent studies of hole transport in a-Si:H have been extended to much thicker films ~30 μm (Mort *et al* 1980b), whereas previously measurements were limited to > 3 μm. The measurements on the thick films have indicated that transit times as long as 1 ms can be observed (Mort *et al* 1980b). This means that the $\mu\tau$ products for holes can be quite comparable to that for electrons. In addition time-of-flight techniques including a delayed-collection field technique have allowed the identification and study of both geminate and

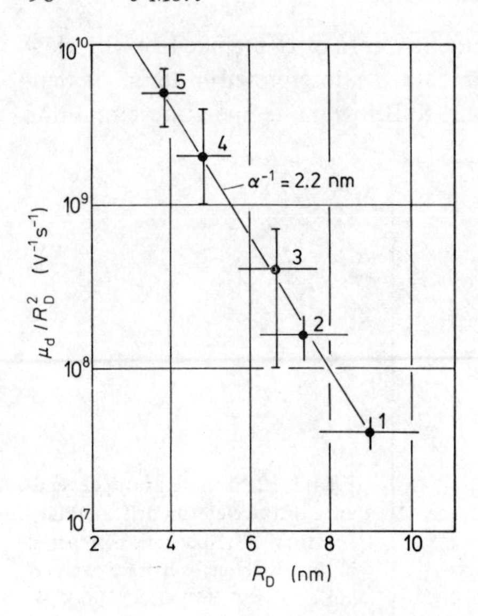

Figure 11. Plot of ln (μ_d/R_D^2) against R_D, where R_D is the average separation between compensated donors in a-Si:H (after Allan *et al* 1977).

non-geminate recombination processes in a-Si (Mort *et al* 1980b). All these results, which have significance for the various technological applications (Chen and Mort 1980), fall outside the scope of this review.

4. Summary

In this necessarily brief review an attempt has been made to give a representative picture of the current experimental status of the study of injected charge transport in dielectric thin films. Key materials and phenomena have been highlighted and similarities and differences in a number of materials discussed. It is clear that considerable progress has been made in characterising transport phenomena by the time-of-flight technique in a broad range of amorphous thin film dielectrics. Similarities exist where the phenomena relate to processes generic to the disordered state. On a more microscopic level the details of the chemical character of the particular material are important and can be discerned in the experimental results.

References

Allan D, LeComber P G and Spear W E 1977 *Proc. 7th Int. Conf. on Amorphous and Liquid Semiconductors* ed. W E Spear (Edinburgh: CICL) p323

Abkowitz M and Enck R C 1980 *J. Non-Cryst. Solids* **35–36** 831

Borsenberger P M, Contois L E and Hoesterey D C 1978 *Chem. Phys. Lett.* **56** 574

Chen I and Mort J 1980 *Appl. Phys. Lett.* in press

Gill W D 1973 *J. Appl. Phys.* **43** 5033

Guttman F and Lyons L E 1967 *Organic Semiconductors* (New York: Wiley)

Kastner M, Adler D and Fritzsche H 1978 *Phys. Rev. Lett.* **37** 1504

Moore A 1977 *Appl. Phys. Lett.* **31** 762

Mort J 1972 *Phys. Rev.* **B5** 3329

— 1980a *Adv. Phys.* **29** 367

— 1980b *Phys. Technol.* **11** 134

Mort J and Chen I 1975 *Appl. Solid State Sci.* **5** 69
Mort J, Chen I, Troup A, Morgan M, Knights J C and Lujan R 1980a *Phys. Rev. Lett.* **45** 1348
Mort J, Grammatica S, Knights J C and Lujan R 1980b *Photogr. Sci. Engng* **24** 241
Mort J and Pfister G 1979 *Polym-Plast. Technol.* **12** 89
Mort J, Pfister G and Grammatica S 1976 *Solid State Commun.* **18** 693
Mott N F, Davis E A and Street R A 1975 *Phil. Mag.* **32** 961
Pfister G 1976 *Phys. Rev. Lett.* **36** 271
— 1977 *Phys. Rev.* **B16** 3676
Pfister G and Griffiths C 1978 *Phys. Rev. Lett.* **40** 659
Pfister G, and Morgan M 1980 *Phil. Mag.* **B41** 209
Pfister G, Mort J and Grammatica S 1976 *Phys. Rev. Lett.* **37** 1360
Pfister G and Scher H 1978 *Adv. Phys.* **27** 747
Ritsko J J and Bigelow R W 1978 *J. Chem. Phys.* **69** 4162
Scharfe M 1970 *Phys. Rev.* **B2** 5025
Scher H and Montroll E W 1975 *Phys. Rev.* **B12** 2455
Slowick J H 1976 *Bull. Am. Phys. Soc.* **21** 314
Spear W E 1957 *Proc. Phys. Soc.* **B70** 669
— 1960 *Proc. Phys. Soc.* **B76** 826
— 1974 *Amorphous and Liquid Semiconductors* eds J Stuke and W Brenig (London: Taylor and Francis) p1
Spear W E and LeComber P G 1972 *J. Non-Cryst. Solids* **8–10** 727
Tabak M D and Hillegas W J 1971 *J. Vac. Sci. Technol.* **9** 387
Tiedje T, Abeles B, Morel D L, Moustakas T D and Wronski C R 1980 *Appl. Phys. Lett.* **36** 695

Inst. Phys. Conf. Ser. No. 58
Invited paper presented at Physics of Dielectric Solids, 8–11 September 1980, Canterbury

Low-mobility transport in perfect crystals

R W Munn
Department of Chemistry, UMIST, Manchester M60 1QD, UK

Abstract. The nature and origins of low-mobility charge transport in perfect dielectric crystals are reviewed. Low mobility arises from weak electron transfer interactions (narrow bands) or strong electron–phonon coupling or both. Transport is often best viewed as involving polarons, quasi-particles consisting of a charge carrier and the associated deformation of the lattice over a small or large distance. Limiting cases of transport are incoherent hopping of carriers between self-trapped states at high temperatures and band motion at low temperatures, but in general the mechanism is not clearly of either type. The development of more general transport theories has been stimulated by work on molecular crystals, where carrier bandwidths, phonon energies, phonon bandwidths, electron–phonon coupling energies and thermal energies may all be comparable. Such theories work from a model Hamiltonian and avoid presuppositions about the nature of transport. A treatment is outlined which encompasses both high-mobility band transport and low-mobility polaron hopping. Carrier coupling to librational modes appears particularly important in certain molecular crystals and recent work on such coupling is discussed. With more penetrating experiments and more realistic theories, a satisfactory understanding of low-mobility transport is being approached.

1. Introduction

Solid dielectrics are poor conductors of electricity. The conductivity tensor for a density n of free carriers of charge $\pm e$ is given by

$$\sigma = n\, e\, \mu, \tag{1}$$

where μ is the carrier mobility tensor; if there are carriers of both sign, each contributes a term like (1). Low conductivity can therefore arise from low carrier densities, low mobilities, or both.

The carrier density can be greatly affected by extrinsic factors such as chemical impurities, physical imperfections and injection at surfaces. In the absence of such factors, the intrinsic thermal equilibrium carrier density is a straightforward matter of statistical thermodynamics: a low density results from a full valence band separated by a wide band gap from an empty conduction band. Wide band gaps commonly occur in crystals composed of closed-shell species such as ionic and van der Waals crystals, in which the mobilities are often also low. This paper therefore considers the carrier mobility in perfect dielectric crystals, where it is the main conduction parameter of interest. Even in this simplest case there are many aspects of low mobility, but a satisfactory understanding is now being approached through a combination of more penetrating experiments and more general theories (Duke and Schein 1980, Roberts *et al* 1980).

0305-2346/81/0058-0098 $01.50

Section 2 deals with some basic principles: what mobility is and when it may be considered low; the various interactions in the crystals which affect the mobility; two simple limiting cases which provide models for transport; and the concept of the polaron as a charge-carrying quasi-particle. Section 3 discusses the problem of calculating the mobility, its temperature dependence and so on, given a knowledge of the interactions which affect it. Only when this problem is solved can one with any confidence attempt the reverse process of deducing the transport mechanism or the nature of the interactions from observed mobility behaviour. The final section summarises the paper and suggests directions in which progress is to be made.

2. Principles

2.1. Mobility

The mobility tensor already introduced relates the drift velocity **v** of the charge carrier to the electric field **E** it experiences, via

$$\mathbf{v} = \boldsymbol{\mu} \cdot \mathbf{E} \,. \tag{2}$$

The electric field actually produces a force which accelerates the carrier until a steady state is reached when dissipative interactions with the lattice reduce the net force to zero; aspects of this time development are discussed by Munn (1974, 1977) and Mott and Stoneham (1977). Carriers do not all have exactly the same velocity, of course, and equation (2) refers to an average velocity, although in amorphous solids, the distribution of velocities may be so wide and asymmetric that the concept of a mobility becomes of limited value (Scher and Montroll 1975, Pfister and Scher 1977). For low enough fields, linear response theory guarantees that μ is independent of **E**. Field-dependent mobilities can be observed in systems containing transition metal ions at fields of about 10^4 V cm^{-1} (Austin and Gamble 1972, Austin and Sayer 1974), though not in anthracene and naphthalene at fields up to 2×10^5 V cm^{-1} (Schein and McGhie 1979b), and so any field dependence is ignored here.

Mobilities can have a wide range of values in different materials. Selected values are shown in table 1 for a range of materials at room temperature. There is no 'natural break' in the sequence, but mobilities below about 10 cm^2 V^{-1} s^{-1} may be regarded as low for present purposes. Such mobilities are conveniently measured by pulse techniques (Spear 1974, Karl 1974).

The mobility tensor is related to the diffusion tensor **D** through the Einstein relation

$$\boldsymbol{\mu} = (e/kT)\, \mathbf{D} \tag{3}$$

This is obtained by equating the conduction and diffusion currents at equilibrium, and comparing the resulting carrier concentration with that required by the Boltzmann distribution. The diffusion tensor may be easier to understand and calculate than the mobility tensor, since it gives the rate of change of the mean square displacement of the carrier. It is however necessary to remember the extra factor T^{-1} in the mobility when temperature dependences are considered.

2.2. Interactions affecting mobility

Carrier mobilities depend on the carrier energy band structure and on factors affecting it. In pure perfect crystals the carrier bands depend only on the instantaneous nuclear

Table 1. Charge-carrier drift mobilities μ for selected solids (at room temperature unless otherwise stated).

Solid	$\mu(cm^2\ V^{-1}\ s^{-1})$	
	Electrons	Holes
Ne (triple point)	600	0.01
Xe (triple point)	4000	0.02
Ge	4000	3000
Si	1500	400
GaAs	8000	300
InSb	77000	1000
NiO	–	4
MnO	–	3×10^{-5}
KCl	10	–
Cu	35	–
Orthorhombic sulphur	5×10^{-4}	5
Anthracene	1.4	1.6

configuration and hence on the phonon band structure. Finally the temperature affects mobilities through the thermal average occupation of the carrier and phonon states.

In low-mobility solids the natural way to view the carrier band structure is in the *tight-binding approximation.* The bands are constructed from the wavefunctions for the isolated ions or molecules with and without a carrier. The band structure is then determined by the carrier *exchange* or *transfer integrals* J between one species and its nearer neighbours.

Coupling of the carrier states to the crystal vibrations is essential to provide the dissipative mechanism required to yield a steady drift velocity and a finite mobility. Viewed another way, this coupling transfers energy from the carrier to the vibrations, giving rise to the familiar phenomena of resistance and Ohmic heating. The coupling may be *local,* affecting the energy of the localised carrier state from which the band is constructed, or *nonlocal,* affecting the transfer integrals (non-Condon behaviour). For example, the conventional deformation-potential coupling gives the energy change in terms of the local elastic strain.

Since most insulators contain compound ions or molecules, their vibrations are complex. Apart from the acoustic modes, the optic modes may contain lattice and molecular vibrations, including librations. Each type of vibration needs to be characterised by (at least) some average frequency ω, a measure of the dispersion or bandwidth Δ, and the strength of the coupling to the carrier states $f\omega$. Normally the dominant coupling will be *linear* in the phonon displacement (the first term in a Taylor series), corresponding to a change of equilibrium position around a carrier. There may in special cases be *quadratic* coupling, corresponding to a change of phonon frequency around a carrier; this may occur when the linear coupling is zero or very small by symmetry (Munn and Siebrand 1970, 1971; Munn and Silbey 1978) or when the transfer integral is very small through the near cancellation of large terms (Tiberghien and Delacôte 1970, Vilfan 1977, Gosar 1980). In practice, the most strongly coupled modes should determine the low mobility and hence are of most interest.

2.3. *Band and hopping transport*

Conventional transport theory uses the *wavevector representation,* which has the advantage of reflecting the translational symmetry of the crystal. On the other hand, in low-mobility solids a *site representation* has advantages, for example in calculating mean square displacements. Treated exactly, either representation gives the same results, but approximations neglecting correlations in one representation imply strong correlations in the other (Fröhlich 1968).

When the electron–phonon coupling is weak, the band states are nearly independent. Carriers can then be regarded as propagating in states of definite wavevector **k** with a velocity $\mathbf{v_k}$, being occasionally scattered by phonons into another state to give a lifetime $\tau_\mathbf{k}$ for each state. The diffusion tensor is then given by

$$\mathbf{D} = \langle \mathbf{v_k v_k} \tau_\mathbf{k} \rangle, \quad (4)$$

where the angle brackets denote a thermal average over the states **k**. As the temperature increases, the number of phonons increases, strengthening the scattering and reducing the lifetimes, so that the band mobility decreases. This is *band transport.* Given a band structure calculation of the $\mathbf{v_k}$, experimental mobilities can be used to deduce a mean lifetime τ. In low-mobility solids the bands are usually narrow (0.1 eV or less) and the velocities correspondingly small, but the mean lifetime may still be small enough to yield an energy uncertainty h/τ comparable with the bandwidth, implying that the band description is inappropriate. Alternatively a mean free path λ may be deduced from the mobility by introducing free paths $\lambda_\mathbf{k} = \tau_\mathbf{k} |\mathbf{v_k}|$ in equation (4). Low mobilities then yield mean free paths comparable with lattice spacings.

In these circumstances, with strong electron–phonon coupling, the site states are nearly independent. Carriers are localised at sites by strong scattering until thermal fluctuations allow a hop to an adjacent site. If the mean hopping frequency is Γ and the nearest neighbour distance is a, the diffusion coefficient is given by

$$D = a^2 \Gamma / 6 \quad (5)$$

in three dimensions. As the temperature increases, the thermal fluctuations increase in amplitude so that the hopping mobility increases, showing an activated behaviour. This is *hopping transport.*

These two limiting cases of band and hopping mobility provide convenient physical pictures of transport, but should not be used as Procrustean beds one or other of which all transport must fit. Mean free paths can clearly take any value from one lattice spacing upwards as the correlation between site states increases and that between band states decreases, so that intermediate types of transport occur. The characteristic temperature dependences in the limiting cases are hardly to be expected when detailed account is taken of all the different couplings described above, and indeed a wide variety of temperature dependences is observed as shown in figure 1, from faster decreases than conventional band theory predicts (Burshtein and Williams 1977, 1978) through almost temperature-independent mobilities (Schein 1977a, b) to exponential activated behaviour (Gibbons and Spear 1966; Loveland *et al* 1972). To rationalise such behaviour, more general theories are needed, treating all the various types of coupling. Approaches of this sort are discussed in the next section.

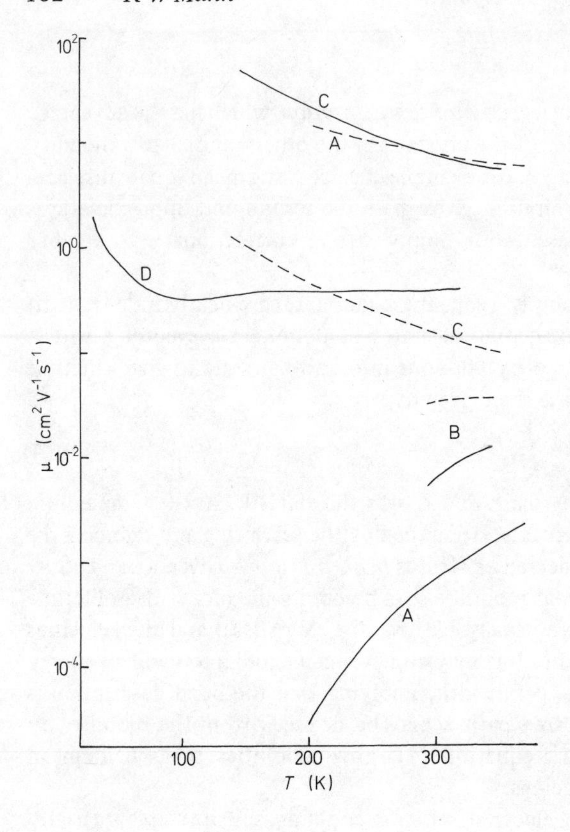

Figure 1. Intrinsic mobilities for selected directions in various crystals as a function of temperature. Full curves: electrons; broken curves: holes. A – orthorhombic sulphur (Adams and Spear 1964, Gibbons and Spear 1966, Ghosh and Spear 1968, Nitzki and Stössel 1970); B – anthracene–1,3,5-trinitrobenzene complex (Zboinski 1976); C – durene, 1,2,4,5-tetramethylbenzene (Burshtein and Williams 1977); D – naphthalene (Schein *et al* 1978, Schein and McGhie 1979a). For comparison, note that the hole mobility in sulphur follows the conventional $T^{-3/2}$ law and the electron mobility in sulphur follows an exponential activation law.

2.4. Polarons

Consider an excess charge carrier $\pm e$ localised in a low-mobility crystal treated as an isotropic dielectric continuum of relative permittivity ϵ_r with a cavity of radius R around the carrier. The carrier energy is lowered through polarisation by

$$P = (e^2/4\pi\epsilon_0 R)(1-1/\epsilon_r). \qquad (6)$$

The high-frequency contribution P_∞ with $\epsilon_r = \epsilon_\infty$ comes from electronic polarisation which adjusts instantaneously as the carrier moves because of the low electronic inertia. The low-frequency or static contribution P_s with $\epsilon_r = \epsilon_s$ comprises not only the electronic polarisation but also nuclear polarisation, which can adjust only if the carrier motion is slow enough. This nuclear polarisation corresponds to an electron–phonon coupling and it contributes an energy

$$\Delta P = P_s - P_\infty = (e^2/4\pi\epsilon_0 R)(1/\epsilon_\infty - 1/\epsilon_s). \qquad (7)$$

When the mobility is low enough, the nuclear displacements move together with the carrier. This increases its inertia and hence its effective mass. The combination of the carrier and its associated lattice deformation is a quasi-particle known as a *polaron.* The particular form of polaron just outlined is the *dielectric* or continuum polaron (Austin and Mott 1969), but similar polarons can be constructed with other forms of electron–phonon coupling. Polarons may also be classified by their range. The dielectric polaron is a *large* polaron, with deformations extending over many lattice spacings through the long-range Coulomb forces. Strong short-range interactions such as valence forces produce a *small* or self-trapped polaron, with deformations confined to a single molecule or unit cell. Examples of a small polaron are a molecular ion with geometry differing from that of the neutral molecule, or a pair of halide ions X^- with an excess hole yielding an X_2^- ion.

The advantage of the polaron concept is that it incorporates a major part of the coupling, leaving a weaker residual coupling between polarons and phonons. This feature is common in theories of strongly interacting systems and recurs in the following section. Polaron transport can then be treated much as ordinary carrier transport but with a new polaron band structure. The main difference is that the polaron is temperature dependent: the polaron bandwidth is a strongly decreasing function of temperature for very strong electron–phonon coupling, leading to a transition from band to hopping transport (Holstein 1959). The transport may also change sharply each time the bandwidth falls by the phonon frequency, as scattering channels involving absorption or emission of $\upsilon + 1$ phonons become energetically unfeasible and only υ phonon processes can lie within the band (Emin 1975, Yarkony and Silbey 1976, 1977).

3. Theories of low mobility

3.1. *Desirable features*

As figure 1 and table 1 illustrate, low mobilities may vary widely in temperature dependence and magnitude. Mobilities may be low for carriers of one sign but high for those of the opposite sign in the same material. As the temperature changes, a mobility may cross the border region between low and high values and its temperature dependence may even change (*cf* naphthalene). Ultimately a theoretical framework is required within which this variety of behaviour can be rationalised and for a given solid related to the interactions discussed in §2.2. As far as possible such a theory must be general in scope, yet tractable so as to yield usable results, and explicit in its relation to the interactions.

A general theory must first avoid prior assumptions about the transport mechanism, so that hopping, band and intermediate types of transport can all be treated on an equal footing. Correspondingly, the mobility must not initially be assumed to be low if the causes of low mobility are to be studied. The theory is thus an extension of conventional theory (to which it should reduce in appropriate limits) to take account of materials in which carrier bandwidths, phonon energies, phonon bandwidths, electron–phonon coupling energies and thermal energies may all be comparable. Molecular crystals are particularly good examples of such materials.

One approach is to study the mean square displacement of a carrier which started at the origin at time t. This is given by

$$\langle R^2(r)\rangle_{\text{av}} = \sum_n R_n^2 G_{nn}(t), \tag{8}$$

where R_n is the distance of site n from the origin and $G_{nn}(t)$ is a diagonal element of the carrier density matrix, giving the probability of finding the carrier at site n at time t. The carrier density matrix $\mathbf{G}(t)$ is obtained from the density matrix $\rho(t)$ for the coupled carrier–phonon system as

$$\mathbf{G}(t) = \text{Tr}_p\,\rho(t). \tag{9}$$

where Tr_p denotes a trace over all phonon variables (Munn and Silbey 1980a). In a crystal, the wavevector representation is usually easier to work in, and then equation (8) becomes

$$\langle R^2(t)\rangle_{\mathrm{av}} = \sum_{\mathbf{k}} \nabla_{\kappa}^2 G_{\mathbf{k},\mathbf{k}+\kappa}(t)|_{\kappa=0}. \tag{10}$$

From the mean square displacement, the diffusion coefficient in three dimensions is given by

$$D = (1/6)\,\mathrm{d}\langle R^2(t)\rangle_{\mathrm{av}}/\mathrm{d}t|_{t\to\infty}, \tag{11}$$

and the mobility follows from equation (3).

Alternatively, in the region of linear response (mobility independent of field) one can use the Kubo formalism (Kubo 1957). The DC conductivity is given by

$$\sigma = (\Omega/2kT)\int_{-\infty}^{\infty} \langle \mathbf{j}(t)\mathbf{j}(0)\rangle\, \mathrm{e}^{-\eta|t|}\,\mathrm{d}t \tag{12}$$

in the limit $\eta\to 0$, where Ω is the volume of the system, and $\mathbf{j}(t)$ is the current density operator in the Heisenberg representation. The angle brackets denote a thermal average over the coupled carrier–phonon system, such that for a matrix A,

$$\langle \mathsf{A}\rangle = \mathrm{Tr}(\rho \mathsf{A}). \tag{13}$$

Both approaches thus depend on the full density matrix $\rho(t)$. Certain exact formal properties of $\rho(t)$ are known, but these are too difficult to apply for realistic systems, and therefore some approximations are required. These amount to neglecting certain couplings or correlations. When all energies can be comparable, such approximations may first require that the Hamiltonian H should be transformed into a new Hamiltonian $\tilde{H}$ with weaker interactions. For example, $\tilde{H}$ may represent a system of polarons and modified phonons rather than bare carriers and phonons. Then $\tilde{H}$ is partitioned in the usual way to separate off a weak interaction V:

$$H \to \tilde{H} = H_0 + V. \tag{14}$$

In particular, when terms up to second order in V are included (in an approximation equivalent to Brillouin–Wigner perturbation theory), the diffusion coefficient (11) can be written as (Yarkony and Silbey 1977, Munn and Silbey 1980)

$$D = \sum_{\mathbf{k}} G_{\mathbf{kk}}(\infty)\,(\mathbf{v}_{\mathbf{k}}^2/\Gamma_{\mathbf{kk}} + \gamma_{\mathbf{kk}}). \tag{15}$$

Here $\mathsf{G}(\infty)$ is the long-time or equilibrium carrier (polaron) density matrix and $\mathbf{v}_{\mathbf{k}}$ is the carrier velocity in state $\mathbf{k}$. The quantity $\Gamma_{\mathbf{kk}}$ is the rate of carrier scattering out of state $\mathbf{k}$, or inverse lifetime $\tau_{\mathbf{k}}^{-1}$, so that the first term in equation (15) looks like the band result (4). The remaining quantity $\gamma_{\mathbf{kk}}$ can be considered as a hopping term, in some limits reducing to $a^2\Gamma_{\mathbf{kk}}$ and so looking like (5).

The $\Gamma_{\mathbf{kk}}$ and $\gamma_{\mathbf{kk}}$ are calculated from quantities

$$W_{\mathbf{k},\mathbf{k}+\mathbf{K};\mathbf{q},\mathbf{q}+\mathbf{K}} = \int_0^{\infty} \mathrm{d}t(\langle V_{\mathbf{q}+\mathbf{K},\mathbf{k}+\mathbf{K}}\, V_{\mathbf{kq}}(t)\rangle_p\, U_{\mathbf{q}+\mathbf{K},\mathbf{k}+\mathbf{K}}(t)$$

$$+ \langle V_{\mathbf{q}+\mathbf{K},\mathbf{k}+\mathbf{K}}(t)\, V_{\mathbf{Kq}}\rangle_p\, U_{\mathbf{kq}}(t)) \tag{16}$$

where $\langle\ldots\rangle_p$ denotes a thermal average over phonon states,

$$U_{\mathbf{kq}}(t) = \exp\left[\mathrm{i}(E_{\mathbf{k}} - E_{\mathbf{q}})t\right], \tag{17}$$

and $V_{\mathbf{kq}}$ is the phonon operator part of V. By using a model Hamiltonian in these expressions, one can relate **D** or μ to the microscopic interactions. Although further approximations may be needed in the evaluation, by starting from a general expression one can explore how different relative values of the parameters in the Hamiltonian lead to different limits of transport behaviour.

3.2. *A general theory*

Recent work on transport theory (Munn and Silbey 1980a, b; Silbey and Munn 1980) exemplifies the sort of approach just discussed. The system is described by the Hamiltonian

$$\begin{aligned} H = & \sum_n \epsilon \mathrm{a}_n^+ \mathrm{a}_n + \sum_{nm} J_{nm} \mathrm{a}_n^+ \mathrm{a}_m \\ & + \sum_{\mathbf{q}} \omega_{\mathbf{q}} (\mathrm{b}_{\mathbf{q}}^+ \mathrm{b}_{\mathbf{q}} + \tfrac{1}{2}) \\ & + N^{-1/2} \sum_{nm\mathbf{q}} \omega_{\mathbf{q}} f_{nm}^{\mathbf{q}} (\mathrm{b}_{\mathbf{q}} + \mathrm{b}_{-\mathbf{q}}^+) \mathrm{a}_n^+ \mathrm{a}_m \end{aligned} \tag{18}$$

where a_n^+ (a_n) creates (destroys) a carrier of energy ϵ at site n and $\mathrm{b}_{\mathbf{q}}^+$ ($\mathrm{b}_{\mathbf{q}}$) creates (destroys) a phonon of energy $\omega_{\mathbf{q}}$ and wavevector $\mathbf{q}$, J_{nm} is the transfer integral between sites n and m, and $f_{nm}^{\mathbf{q}}$ is the linear electron–phonon coupling parameter (local for n=m, nonlocal otherwise). In the wavevector representation for the carriers the Hamiltonian becomes

$$\begin{aligned} H = & \sum_{\mathbf{k}} \epsilon_{\mathbf{k}} \mathrm{a}_{\mathbf{k}}^+ \mathrm{a}_{\mathbf{k}} + \sum_{\mathbf{q}} \omega_{\mathbf{q}} (\mathrm{b}_{\mathbf{q}}^+ \mathrm{b}_{\mathbf{q}} + \tfrac{1}{2}) \\ & + N^{-1/2} \sum_{\mathbf{kq}} \omega_{\mathbf{q}} f_{\mathbf{k}}^{\mathbf{q}} (\mathrm{b}_{\mathbf{q}} + \mathrm{b}_{-\mathbf{q}}^+) \mathrm{a}_{\mathbf{k}+\mathbf{q}}^+ \mathrm{a}_{\mathbf{k}}, \end{aligned} \tag{19}$$

where $\epsilon_{\mathbf{k}} = \epsilon + J_{\mathbf{k}}$, with $J_{\mathbf{k}}$ and $f_{\mathbf{k}}^{\mathbf{q}}$ the Fourier transforms of J_{nm} and $f_{nm}^{\mathbf{q}}$.

The Hamiltonian is subjected to the generalised polaron transformation (Munn and Silbey 1980b)

$$H \to \tilde{H} = e^S H e^{-S}, \tag{20}$$

where

$$S = N^{-1/2} \sum_{nm\mathbf{q}} A_{nm}^{\mathbf{q}} (\mathrm{b}_{-\mathbf{q}}^+ - \mathrm{b}_{\mathbf{q}}) \mathrm{a}_n^+ \mathrm{a}_m, \tag{21}$$

with $A_{nm}^{\mathbf{q}}$ to be determined in some optimal way. This transformation displaces the equilibrium positions for the phonons (i.e. distorts the lattice) in a manner dependent on the carrier distribution. It also mixes the carrier state on one site with a superposition of carrier states on nearby sites through operators θ_{nm} depending on the phonon momenta, to yield polaron operators:

$$\mathrm{a}_n \to \sum_m \theta_{nm} \mathrm{a}_m, \tag{22}$$

$$\theta_{nm} = \{\exp [N^{-1/2} \sum_{\mathbf{q}} \mathbf{A}^{\mathbf{q}}(b_{\mathbf{q}} - b^{+}_{-\mathbf{q}})]\}_{nm}, \tag{23}$$

where $\mathbf{A}^{\mathbf{q}}$ is the matrix of the $A^{\mathbf{q}}_{nm}$.

Where energies are concerned, the best transformation should be determined variationally to minimise the free energy (Yarkony and Silbey 1976, 1977). For transport, a less rigorous method of choice may be expedient to make the residual coupling V tractable. Here the original coupling term in H is eliminated in the average over the thermal equilibrium free phonon ensemble by setting

$$A^{\mathbf{q}}_{rs} = \sum_{nm} f^{\mathbf{q}}_{nm} \langle \theta^{+}_{nr} \theta_{ms} \rangle_{p}. \tag{24}$$

Since θ_{nm} depends on $\mathbf{A}^{\mathbf{q}}$, this expression determines the $A^{\mathbf{q}}_{rs}$ only implicitly. The transformed Hamiltonian $\tilde{H}$ is then partitioned so that V consists of the deviation of the carrier–phonon coupling from its thermal phonon average. This ensures that at long times the proper equilibrium state is attained (Silbey 1976).

With these choices, the zeroth-order Hamiltonian becomes

$$H_0 = \sum_{\mathbf{k}} \tilde{\epsilon}_{\mathbf{k}} a^{+}_{\mathbf{k}} a_{\mathbf{k}} + \sum_{\mathbf{q}} \omega_{\mathbf{q}} (b^{+}_{\mathbf{q}} b_{\mathbf{q}} + \tfrac{1}{2}) \tag{25}$$

where the polaron energy $\tilde{\epsilon}_{\mathbf{k}}$ is given by

$$\tilde{\epsilon}_{\mathbf{k}} = \epsilon + \tilde{J}_{\mathbf{k}} - N^{-1} \sum_{\mathbf{q}} \omega_{\mathbf{q}} |A^{\mathbf{q}}_{\mathbf{k}}|^2. \tag{26}$$

This differs from the carrier energy $\epsilon_{\mathbf{k}}$ in two ways. First, the wavevector-dependent part $J_{\mathbf{k}}$ has been renormalised to $\tilde{J}_{\mathbf{k}}$, the Fourier transform of

$$\tilde{J}_{rs} = \sum_{nm} J_{nm} \langle \theta^{+}_{nr} \theta_{ms} \rangle_{p}. \tag{27}$$

Here the thermal average is essentially an exponential of negative argument, which therefore narrows the polaron band compared with the free carrier band. The narrowing is more marked as the temperature increases and as the local coupling increases, but stronger nonlocal coupling tends to offset the narrowing by increasing the carrier kinetic energy (Umehara 1979). Secondly, all the energies are lowered by the polaron binding energy depending on $A^{\mathbf{q}}_{\mathbf{k}}$, the Fourier transform of $A^{\mathbf{q}}_{rs}$. Nonlocal coupling makes the binding energy dependent on $\mathbf{k}$, thereby altering the band shape. These changes are illustrated in figure 2.

The residual coupling V takes a rather complicated form containing as a factor the combination

$$T_{nt,ms} = \theta^{+}_{nt} \theta_{ms} - \langle \theta^{+}_{nt} \theta_{ms} \rangle_{p}, \tag{28}$$

so that $\langle V \rangle_p$ is clearly zero. This combination ensures that V is small as required. For weak coupling the θ operators differ from the unit operator and hence from their thermal average only by small terms, making $T_{nt,ms}$ and the second-order averages in equation (16) correspondingly small, while for strong coupling the averages become exponentials of large negative argument (like those causing the band narrowing) and again small. The form of V therefore ensures a smooth variation between the two limits without divergences characteristic of the breakdown of perturbation theory.

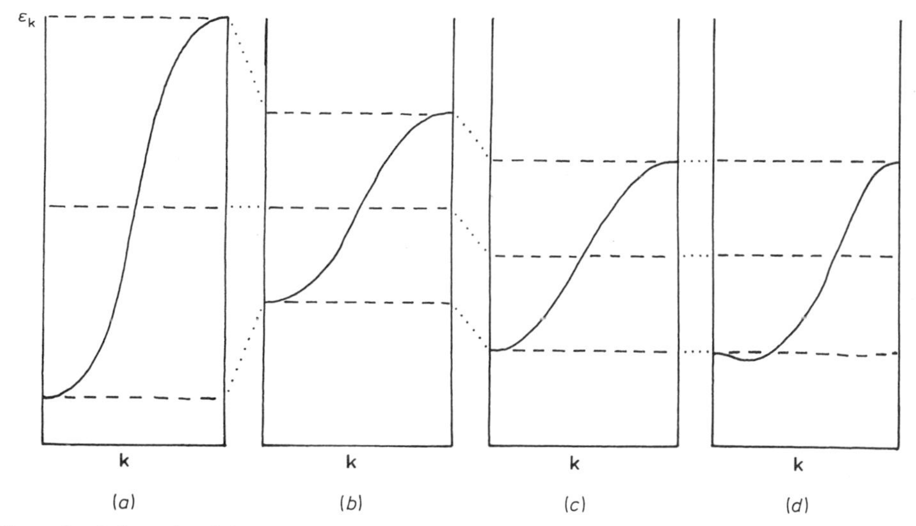

Figure 2. Polaron band formation (schematic). (*a*) Bare carrier band. (*b*) Band narrowing. (*c*) Band lowering. (*d*) Change of band shape.

Transport in this model has so far been examined only for local electron–phonon coupling (Silbey and Munn 1980). This is simpler than nonlocal coupling but includes such standard types as the deformation-potential coupling for acoustic modes and dielectric and short-range coupling to lattice and molecular optic modes (some aspects of transport with nonlocal coupling are discussed in §3.3). The carrier bandwidth is taken as B, reducing to $\tilde{B}$ for the polaron. The strength of the electron–phonon coupling is described by the dimensionless parameter g.

For weak coupling, the last term in equation (18) can be used as the perturbation without transformation; as required, the results can also be obtained as the weak-coupling limit of the transformed results. In this limit, scattering is weak and hopping negligible, so that mobilities tend to be high. Table 2 shows how the mobility depends on the carrier bandwidth B ($\approx \tilde{B}$ for weak coupling), the temperature T and the coupling strength g for scattering (a) by a Debye spectrum of acoustic phonons, maximum frequency ω_D, assuming deformation-potential coupling (where $g\omega_D$ may be in turn proportional to B),

Table 2. Form of the mobility for weak local electron–phonon coupling in various limits of the other parameters.

Limit		Form of μ
	(a) Acoustic phonon scattering	
$B \gg kT \gg \omega_D$		$B^{5/2}/g^2\omega_D T^{3/2}$
$kT \gg B \gg \omega_D$		$B^3/g^2\omega_D T^2$
$kT \gg \omega_D \gg B$		$\omega_D^2/g^2 T^2$
	(b) Optic phonon scattering	
$B \gg \omega \gg kT$		$B^{5/2}/g^2\omega^2 T^{1/2}$
$B \gg kT \gg \omega$		$B^{5/2}/g^2\omega T^{3/2}$
$kT \gg \omega \gg B$		$B\omega^2/g^4 T^3$

and (b) by a narrow band of optic phonons, mean frequency ω. For acoustic phonons, and for optic phonons when $B > \omega$, one-phonon scattering is possible, so that $\Gamma_{\mathbf{kk}} \sim g^2$ and from equation (15) $\mu \sim 1/g^2$. For optic phonons when $B < \omega$, at least two phonons are required for scattering and $\mu \sim 1/g^4$. The wide-band high-temperature results show the standard temperature dependence $\mu \sim T^{-3/2}$, but for narrow bands μ varies as T^{-2} (Glarum 1963, Friedman 1965) or even as T^{-3}. Even when other types of scattering are considered, μ always decreases with increasing temperature as T^{-1} or faster for narrow band transport (Schein 1977a).

For strong coupling, the transformed Hamiltonian must be used. The phonons are assumed to be optic modes (lattice or molecular) with a mean frequency ω and a bandwidth Δ much less than ω. For analytical convenience the phonon and polaron bands are both assumed to have a gaussian density of states, the polaron bandwidth being

$$\tilde{B} = B \exp\left[-(2v + 1) g^2\right], \tag{29}$$

where

$$v = 1/\left[\exp(\omega/kT) - 1\right] \tag{30}$$

is the thermal average number of phonons of frequency ω. The averages in equation (16) show a complicated decaying oscillatory behaviour as a function of time. This decay is approximately gaussian of width Γ, where

$$\Gamma = \Delta \qquad (y \leqslant 1), \tag{31}$$

$$\Gamma = \Delta y^{1/2} \qquad (y > 1), \tag{32}$$

with

$$y = 4g^2 \left[v(v+1)\right]^{1/2}. \tag{33}$$

Under these conditions, with the neglect of minor terms the hopping term $\gamma_{\mathbf{kk}}$ reduces to $a^2\Gamma_{\mathbf{kk}}$: the site hopping rate equals the band scattering rate (Holstein 1959). When single-phonon scattering is prevented by setting $B \leqslant \omega$, the scattering rate becomes

$$\Gamma_{\mathbf{kk}} = \frac{\pi^{1/2}\tilde{B}^2\left[I_0(y) - 1\right]}{4(\tilde{B}^2 + \Gamma^2)^{1/2}} \exp\left[-\frac{E_{\mathbf{k}}^2}{\tilde{B}^2 + \Gamma^2}\right] \tag{34}$$

where $I_0(y)$ is the modified Bessel function of order zero. Here the combination $\tilde{B}^2 + \Gamma^2$ describes a combined electronic and vibrational relaxation or dephasing rate, with no assumptions made about the relative sizes of $\tilde{B}$ and Γ. The mobility obtained from equation (34) is given by

$$\frac{\mu kTh}{ea^2} = \frac{(\tilde{B}^2 + \Gamma^2)}{\pi^{1/2}(\tilde{B}^2 + 2\Gamma^2)^{1/2}} \frac{1}{\left[I_0(y) - 1\right]} + \frac{\pi^{1/2}\tilde{B}^2\left[I_0(y) - 1\right]}{4(2\tilde{B}^2 + \Gamma^2)^{1/2}} \exp\left[-\frac{\tilde{B}^4}{4k^2T^2(2\tilde{B}^2 + \Gamma^2)}\right] \tag{35}$$

where the first term represents the band contribution, the second the hopping contribution. The detailed behaviour of this expression in different limits is discussed elsewhere (Silbey and Munn 1980).

The mobility obtained from equation (35) is shown as a function of temperature for various coupling strengths in figure 3. The bare carrier bandwidth B is taken to be equal to ω and the phonon bandwidth Δ is taken as $0.1\ \omega$. The mobilities are expressed as multiples of ea^2/h, which for $a^2 = 50 \times 10^{-20}\ \text{m}^2$ corresponds to $1.2\ \text{cm}^2\ \text{V}^{-1}\ \text{s}^{-1}$. As g increases, the curves change in shape because the band term decreases more rapidly with increasing temperature, the hopping term increases more rapidly, and the maximum in the hopping term changes position in a complex way (accounting for the apparently anomalous behaviour of the curve for $g^2 = 0.3$). For the largest coupling strengths, the band term decreases so rapidly and the hopping term increases so rapidly that they are comparable only in a very narrow temperature range. This produces a minimum in the mobility which can be considered to mark a transition from pure band to pure hopping transport (Holstein 1959). For smaller carrier bandwidths, the mobility decreases and a stronger coupling is required to produce the minimum. For smaller phonon bandwidths, the band term decreases and the hopping term increases.

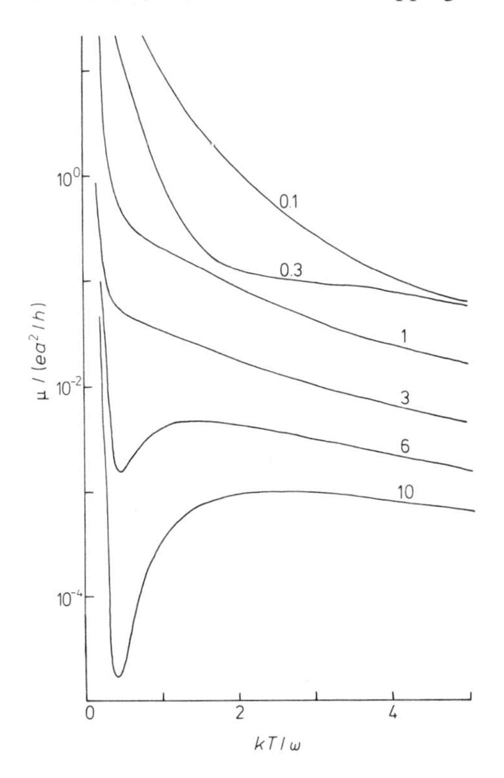

Figure 3. Mobility as a function of temperature for $B = \omega$, $\Delta = 0.1\ \omega$. Each curve is marked with the appropriate value of g^2.

Comparison of figures 1 and 3 suggests that equation (35) is capable of providing a reasonable account of the temperature dependence of many mobilities, particularly when it is realised that, for reasonable values of the frequency ω, figure 3 covers a much wider temperature range than the data in figure 1 (double the range, if $\omega = 100\ \text{cm}^{-1}$). However, the results shown in figure 1 for electrons in the c' direction of naphthalene are not easily explained. The rise in mobility below 100 K may be attributed to the onset of band transport, but figure 3 shows no indication of such a large region of almost constant mobility. Such regions are also found for electrons in the b direction in naphthalene, in the b and c' directions in deuterated naphthalene, in the c' direction in anthracene and deuterated anthracene, and perpendicular to the layers in arsenic(III) sulphide (As_2S_3, orpiment) (Schein 1977a, b, Schein and McGhie 1979a, Duke and Schein 1980). More specialised theories have been developed to explain this behaviour, as described in the following subsection.

3.3. Special theories

The almost temperature-independent mobilities are observed in the directions perpendicular to the crystal cleavage planes, except for the naphthalene b direction. These directions correspond to weak interactions, and in anthracene and naphthalene the transfer integrals in these directions are calculated to be small (Tiberghien and Delacôte 1970) – so small, indeed, that the mobilities would be expected to be much smaller than observed. The calculations also show that J is so small because of the near cancellation of two contributions in the equilibrium crystal structure, but that J may increase by a factor of 5 or 10 for plausible values of the molecular rotation (Tiberghien and Delacôte 1970). This implies a very strong nonlocal electron–phonon coupling $f^{\mathbf{q}}_{mn}$ for librational modes $\mathbf{q}$ relating molecules m and n in adjacent planes.

Nonlocal coupling produces an extra *phonon-assisted* contribution to the current density $\mathbf{j}$, thereby increasing the mobility, as the Kubo formula (12) indicates (Gosar and Choi 1966, Gosar and Vilfan 1970, Vilfan 1973). Calculation of the extra mobility contribution showed that it was an order of magnitude larger than the usual contribution in anthracene, with almost exactly the opposite temperature dependence (Vilfan 1977). However, this work did not fully take into account the different nature of the transport within and between the planes.

Other work treats the transport as hopping between the planes and band motion within them (Sumi 1979a), obtaining a temperature dependence for each direction resembling that observed in anthracene and naphthalene (except for the b direction). Extensions of this work (Sumi 1979b) show how defect scattering may affect the region of the shallow minimum, which could explain differences observed by different groups in this region (Warta 1978, Karl 1980). The high electric fields used in such experiments at low temperatures produce energy differences between crystal planes of the same order of magnitude as thermal and phonon energies, and must also be taken into account. These calculations predict nonlinear transport under certain conditions (Sumi 1979b), but the mobility remains constant to the highest fields so far applied (Schein and McGhie 1979b). This approach has been extended to include exponential rather than linear coupling in layer crystals (Gosar 1980). This can offset the normal fall in the mobility by making the mean transfer integral $\langle J \rangle_p$ increase with increasing temperature; for linear coupling $\langle J \rangle_p$ is constant.

4. Conclusions

Low-mobility transport is very common and very complicated, even in perfect crystals. Mobilities can be low in various types of crystal, but it is in molecular crystals that conventional assumptions and theories of the solid state best reveal their limitations. It has indeed been asserted that transport results in molecular crystals ‘challenge severely both traditional band models and hopping models, thereby constituting one of the major outstanding mysteries in solid-state physics' (Duke and Schein 1980). Improved models are required to rationalise observed low-mobility transport in terms of crystal structure and interactions, and ultimately to predict mobilities with some reliability.

Successful models must be less restrictive than hitherto in making assumptions about mechanisms of transport and the relative importance of interactions. In more general models like this, it is high mobility which is the special case rather than low mobility. Progress in this direction has been made with the development of a theoretical framework encompassing both conventional semiconductor mobility theory and small-polaron mobility theory (Silbey and Munn 1980). Though limited to low carrier densities, this theory is able to treat carrier bands of arbitrary width and hence should be valuable for crystals with highly anisotropic band structures. (An approach along similar lines might perhaps also be useful in amorphous solids, but the absence of translational symmetry and the need for configurational averaging present obvious difficulties.)

However, a more searching test of theory is provided by new experiments using higher electric fields over wider temperature ranges on ultra-pure samples (Schein 1977b, Schein and McGhie 1979a, Warta 1978, Karl 1980). These have stimulated the development of more specialised theories treating nonlocal coupling in anisotropic crystals (Vilfan 1977, Sumi 1977a, b, Gosar 1980). The next stage is to incorporate these theories in a more general framework, for which the lines of development can already be foreseen. More effort will then be required in the calculation or experimental estimation of carrier transfer integrals (band structure) and electron–phonon coupling constants, so that quantitative interpretation of low mobility can proceed in earnest, stimulated as always by high-quality experimental measurements.

Acknowledgment

The work in §3.2 was performed in collaboration with Professor R Silbey, with support from NATO grant 1054.

Note added in proof. A field-dependent mobility has recently been reported for electrons in the c' direction of anthracene for fields above about 1.5×10^5 V cm^{-1} at 140 K; no field dependence is observed up to 3×10^5 V cm^{-1} at 290 K (Nakano S and Maruyama Y 1980 *Solid State Commun.* **35** 671–3).

References

Adams A R and Spear W E 1964 *J. Phys. Chem. Solids* **25** 1113–38
Austin I G and Gamble R 1972 *Conduction in Low-Mobility Materials* (London: Taylor and Francis) pp1–17
Austin I G and Mott N F 1969 *Adv. Phys.* **18** 41–102
Austin I G and Sayer M 1974 *J. Phys. C: Solid State Phys.* **7** 905–24
Burshtein Z and Williams D F 1977 *Phys. Rev.* **B15** 5769–79
— 1978 *J. Chem. Phys.* **68** 983–8
Duke C B and Schein L B 1980 *Phys. Today* **33(2)** 42–8
Emin D 1975 *Adv. Phys.* **24** 305–48
Friedman L 1965 *Phys. Rev.* **140** A1649–67
Fröhlich H 1968 *Helv. Phys. Acta* **41** 838–9

Ghosh P K and Spear W E 1968 *J. Phys. C: Solid State Phys.* **1** 1347–58
Gibbons D J and Spear W E 1966 *J. Phys. Chem. Solids* **27** 1917–25
Glarum S H 1963 *J. Phys. Chem. Solids* **24** 1577–83
Gosar P 1971 *Phys. Rev.* **B3** 1991–9
— 1980 *Proc. Ann. Conf. Condensed Matter Division EPS, Antwerp, Belgium,* to be published
Gosar P and Choi S–i 1966 *Phys. Rev.* **150** 529–38
Gosar P and Vilfan I 1970 *Mol. Phys.* **18** 49–61
Holstein T 1959 *Ann. Phys., NY* **8** 343–89
Karl N 1974 *Festkörperprobleme* **14** 261–90
— 1980 *Ninth Molecular Crystal Symposium, Kleinwalsertal, Austria*
Kubo R 1957 *J. Phys. Soc. Japan* **12** 570–86
Loveland R J, LeComber P G and Spear W E 1972 *Phys. Rev.* **B6** 3121–7
Mott N F and Stoneham A M 1977 *J. Phys. C: Solid State Phys.* **10** 3391–8
Munn R W 1974 *Chem. Phys.* **6** 469–73
— 1977 *Chem. Phys. Lett.* **52** 168–70
Munn R W and Siebrand W 1970 *J. Chem. Phys.* **52** 6391–406
— 1971 *Disc. Faraday Soc.* **51** 17–23
Munn R W and Silbey R 1978 *J. Chem. Phys.* **68** 2439–50
— 1980a *Mol. Cryst. Liq. Cryst.* **57** 131–44
— 1980b *Ninth Molecular Crystal Symposium, Kleinwalsertal, Austria*
Nitzki V and Stössel W 1970 *Phys. Status Solidi* **39** 309–16
Pfister G and Scher H 1977 *Phys. Rev.* **B15** 2062–83
Roberts G G, Apsley N and Munn R W 1980 *Phys. Rep.* **60** 59–150
Schein L B 1977a *Phys. Rev.* **B15** 1024–34
— 1977b *Chem. Phys. Lett.* **48** 571–4
Schein L B, Duke C B and McGhie A R 1978 *Phys. Rev. Lett.* **40** 197–200
Schein L B and McGhie A R 1979a *Phys. Rev.* **B20** 1631–9
— 1979b *Chem. Phys. Lett.* **62** 356–9
Scher H and Montroll E W 1975 *Phys. Rev.* **B12** 2455–77
Silbey R 1976 *Ann. Rev. Phys. Chem.* **27** 203–23
Silbey R and Munn R W 1980 *J. Chem. Phys.* **72** 2763–73
Spear W E 1974 *Adv. Phys.* **23** 523–46
Sumi H 1979a *J. Chem. Phys.* **70** 3775–85
— 1979b *J. Chem. Phys.* **71** 3403–11
Tiberghien A and Delacote G 1970 *J. Physique* **31** 637–56
Umehara M 1979 *J. Phys. Soc. Japan* **47** 852–60
Vilfan I 1973 *Phys. Status Solidi* b **59** 351–60
— 1977 *Lecture Notes in Physics* **65** 629–36
Warta W 1978 *Diplomarbeit* University of Stuttgart
Yarkony D R and Silbey R 1976 *J. Chem. Phys.* **65** 1042–52
— 1977 *J. Chem. Phys.* **67** 5818–27
Zboiński Z 1976 *Phys. Status Solidi* b **74** 561–6

Inst. Phys. Conf. Ser. No. 58
Invited paper presented at Physics of Dielectric Solids, 8–11 September 1980, Canterbury

Dispersion forces

A D Buckingham
University Chemical Laboratory, Lensfield Road, Cambridge, CB2 1EW

Abstract. The origin of the long-range attractive forces between inert gas atoms is considered. The general perturbation theory of long-range intermolecular forces is reviewed and formulae for the dispersion energy derived. Estimates of the magnitudes of the R^{-6}, R^{-8} and R^{-10} contributions to the dispersion energy for inert gas and alkali metal atoms are given. Non-additivity of dispersion interactions, and forces between macroscopic bodies, are briefly discussed.

1. Introduction

Atoms and molecules must attract one another when they are far apart, since liquids and solids exist. Likewise they must repel one another when close, since the densities of liquids and solids have their observed values. Figure 1 illustrates this important truth and shows a typical interaction energy $U(R)$ for two spherical atoms as a function of their separation R. For two argon atoms, the well depth ϵ is 0.198×10^{-20} J ($\epsilon k = 143$ K) and the equilibrium separation R_e is 3.76×10^{-10} m (Barker 1976).

The attractive force between two inert gas atoms is due entirely to the *dispersion forces.* These are quantum mechanical in origin – being due to the zero-point motion of the electrons – and were given this name by London (1937) because they have a similar origin to the well known frequency dependence of the response of a single atom to electromagnetic radiation in the visible or ultraviolet part of the spectrum. They are also sometimes known as London forces.

The dispersion interaction energy varies as R^{-6} at large R, giving a force proportional to R^{-7}. It is said to be 'long-range' in contrast to 'short-range' interaction which varies

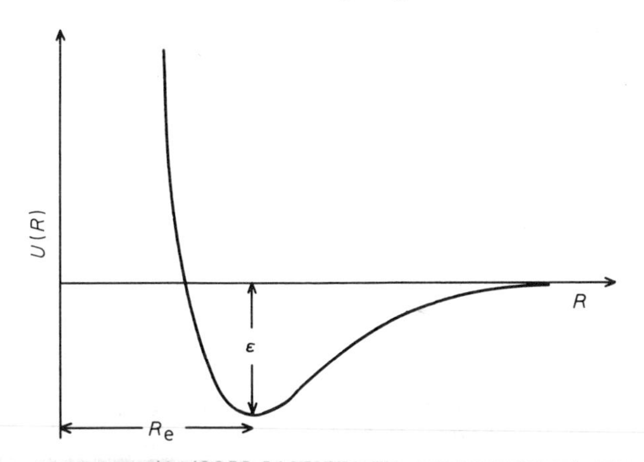

Figure 1. The interaction energy $U(R)$ of two inert gas atoms as a function of their separation.

0305-2346/81/0058-0113 $01.50

exponentially with R, for example as $\exp(-aR)$. Short-range interactions result from electron overlap. The repulsion at separations smaller than R_e in figure 1 may be attributed to the coulomb and exchange interactions; unlike the dispersion forces, these short-range forces may be described quite well by a one-electron theory. Thus standard self-consistent field computations yield reasonable repulsive potentials. But dispersion forces are absent in a one-electron theory, and they are due to interatomic electron correlation.

The magnitude of dispersion forces may be comparable to the electrostatic forces between dipolar molecules. Thus the energy of sublimation of argon at 0 K is 7.74 kJ mol^{-1} (Barker 1976) and this is approximately 40% of that of the isoelectronic polar species HCl, which has a permanent dipole moment of 1.11 D = 3.70×10^{-30} C m. It is hardly surprising that the origin of the attractive forces in the case of the inert gases was obscure before the advent of quantum mechanics. Debye (1921) even made the suggestion that these atoms may possess permanent quadrupole moments!

It is now possible to use atomic spectral data to calculate long-range dispersion interaction energies to high accuracy (~1%) (Dalgarno 1967). There have been significant developments too in the evaluation of forces between macroscopic bodies (Lifshitz 1956, McLachlan 1965, Langbein 1974, Richmond 1975, Mahanty and Ninham 1976). These advances relate the dispersion forces to the frequency-dependent dielectric constants of the samples and permit calculations to be performed for macroscopic bodies of simple shape; retardation effects may be incorporated in these calculations without difficulty.

2. A pair of coupled oscillators

Since *understanding* is stressed in this conference, we begin our detailed discussion of long-range forces by consideration of a very simple system – two identical one-dimensional coupled oscillators, each of mass m and force constant K (figure 2).

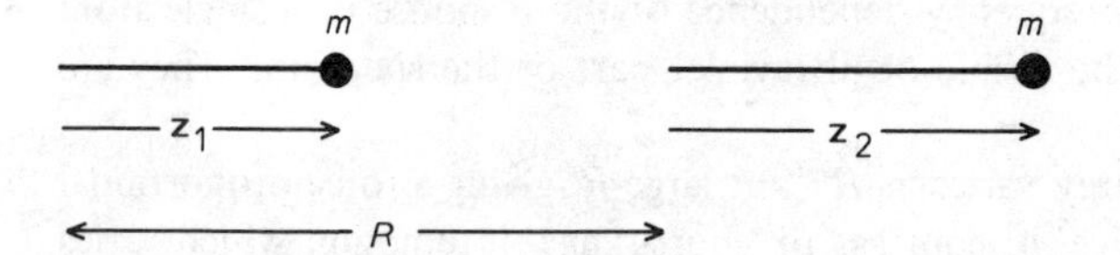

Figure 2. A pair of identical linear oscillators. The interaction potential energy is proportional to $z_1 z_2$.

The Hamiltonian for each free oscillator is $(1/2m)\, p_z^2 + (1/2)Kz^2$ where p_z is the momentum operator and z the displacement from equilibrium. For the coupled pair

$$\mathscr{H} = \frac{1}{2m}(p_{1z}^2 + p_{2z}^2) + \tfrac{1}{2}K(z_1^2 + z_2^2 + 2Cz_1 z_2). \tag{1}$$

If the oscillators are considered to be electric dipoles of moments qz_1 and qz_2, the coupling constant C is $-2(q^2/KR^3)(4\pi\epsilon_0)^{-1}$.

Equation (1) can be re-arranged to a separated form:

$$\mathscr{H} = \frac{1}{4m}[(p_{1z} + p_{2z})^2 + (p_{1z} - p_{2z})^2]$$
$$+ \tfrac{1}{4}K[(1 + C)(z_1 + z_2)^2 + (1 - C)(z_1 - z_2)^2]. \tag{2}$$

The oscillations therefore split into two modes of angular frequency ω_+ and ω_- where

$$\omega_\pm = \omega_0 (1 \pm C)^{1/2}, \qquad \omega_0 = (K/m)^{1/2}. \tag{3}$$

If one uses classical statistical mechanics to evaluate the thermodynamic functions of a system of coupled oscillators of frequency $\omega_\pm$ (Longuet-Higgins 1965), one finds that the free energy of interaction of the oscillators is purely entropic, the change in internal energy being zero. But in reality, the oscillators possess zero-point energy which changes on coupling. For the ground state

$$U = \tfrac{1}{2}\hbar(\omega_+ + \omega_- - 2\omega_0) = \tfrac{1}{2}\hbar\omega_0 \,[(1+C)^{1/2} + (1-C)^{1/2} - 2]. \tag{4}$$

This is negative for positive or negative C and vanishes when C is zero. For small C

$$U = -\tfrac{1}{8}\hbar\omega_0 \,[C^2 + \tfrac{5}{16}C^4 + \ldots]. \tag{5}$$

Thus the interaction energy varies *quadratically* with the strength of the coupling. This is the dispersion energy and it arises from the correlation in the motion of the quantum oscillators. For the coupled dipole oscillators, C is proportional to R^{-3} and U varies as R^{-6} at large R.

If there is a vibrational quantum in one of the identical oscillators, the coupling lifts the degeneracy and gives a resonance energy linear in C. The energy levels are

$$U'_+ = \tfrac{3}{2}\hbar\omega_+ + \tfrac{1}{2}\hbar\omega_- = \hbar\omega_0 \,[\tfrac{3}{2}(1+C)^{1/2} + \tfrac{1}{2}(1-C)^{1/2}], \tag{6}$$

$$U'_- = \tfrac{1}{2}\hbar\omega_+ + \tfrac{3}{2}\hbar\omega_- = \hbar\omega_0 \,[\tfrac{1}{2}(1+C)^{1/2} + \tfrac{3}{2}(1-C)^{1/2}]. \tag{7}$$

The mean interaction energy $\tfrac{1}{2}(U'_- + U'_+) - 2\hbar\omega_0$ is twice that in the ground state. There is a splitting

$$U'_- - U'_+ = \hbar\omega_0 \,[(1-C)^{1/2} - (1+C)^{1/2}] \tag{8}$$

and for small C this reduces to $-\hbar\omega_0 [C + \tfrac{1}{8}C^3 + \ldots]$. The resonance energy is $\pm\tfrac{1}{2}(U'_- - U'_+)$; it results from the lifting of the degeneracy of the excited states by the coupling.

3. General theory of long-range two-body forces

Consider the interaction of the molecule a in the electronic state m_a with molecule b in state m_b. The effects of electron exchange between a and b are neglected, so the treatment is limited to the long-range regime. The effects of exchange lead to interactions which decreases exponentially with R at large R. The Hamiltonian is

$$\mathcal{H} = \mathcal{H}^{(a)} + \mathcal{H}^{(b)} + \mathcal{H}' \tag{9}$$

where $\mathcal{H}^{(a)}$ and $\mathcal{H}^{(b)}$ are the clamped-nuclei Hamiltonians of the isolated molecules and $\mathcal{H}'$ is the interaction:

$$\mathcal{H}' = (4\pi\epsilon_0)^{-1} \sum_{i,i'} e_i^{(a)} e_{i'}^{(b)} R_{ii'}^{-1} \tag{10}$$

where $R_{ii'}$ is the separation of the ith charge $e_i^{(a)}$ in molecule a from the i'th charge in b. A general theory of long-range forces is obtained by treating $\mathcal{H}'$ as a perturbation to

$\mathscr{H}^{(a)} + \mathscr{H}^{(b)}$. The unperturbed wavefunctions are simple products $p_a p_b$ of the eigenfunctions of $\mathscr{H}^{(a)}$ and $\mathscr{H}^{(b)}$.

Using elementary perturbation theory the interaction energy of the pair is

$$U_{m_a m_b} = \langle m_a m_b | \mathscr{H}' | m_a m_b \rangle - \sum_{p_a p_b}{}' \frac{|\langle p_a p_b | \mathscr{H}' | m_a m_b \rangle|^2}{W_{p_a} - W_{m_a} + W_{p_b} - W_{m_b}} + \ldots \tag{11}$$

where Σ' denotes a sum over all unperturbed states except $m_a m_b$, and $W_{p_a} - W_{m_a}$ is the energy difference between unperturbed states of a. If $m_a m_b$ is degenerate, and if this degeneracy is lifted by $\mathscr{H}'$, the zero-order wavefunction $m_a m_b$ is chosen so that $\mathscr{H}$ is diagonal. The first-order energy $\langle m_a m_b | \mathscr{H}' | m_a m_b \rangle$ is the electrostatic energy together with the resonance energy in the degenerate case.

The second-order energy in (11) may be separated into two types:

$$U^{(2)}_{m_a m_b} = (U^{(a)}_{\text{induction}} + U^{(b)}_{\text{induction}}) + U_{\text{dispersion}} \tag{12}$$

where

$$U^{(a)}_{\text{induction}} = -\sum_{p_a \neq m_a} \frac{|\langle p_a m_b | \mathscr{H}' | m_a m_b \rangle|^2}{W_{p_a} - W_{m_a}} \tag{13}$$

$$U_{\text{dispersion}} = -\sum_{\substack{p_a \neq m_a \\ p_b \neq m_b}} \frac{|\langle p_a p_b | \mathscr{H}' | m_a m_b \rangle|^2}{W_{p_a} - W_{m_a} + W_{p_b} - W_{m_b}} \tag{14}$$

The dispersion energy results from matrix elements of $\mathscr{H}'$ which are off-diagonal in the electronic eigenfunctions of both molecules, that is, from correlation in the fluctuations in the charge distributions of different molecules. Like the induction energy (which results from the distortion of a molecule by electric fields arising from its neighbours), it is negative when both molecules are in their ground states.

We require a theory that expresses the long-range interaction energy in terms of properties of the free molecules. This may be achieved by expanding $\mathscr{H}'$ in a multipole series (Buckingham 1978):

$$\mathscr{H}' = \sum_{n=0}^{\infty} \sum_{n'=0}^{\infty} (-1)^n \frac{2^n n!}{(2n)!} \frac{2^{n'} n'!}{(2n')!} \times T_{\alpha\beta\ldots\nu\alpha'\beta'\ldots\nu'} \, \xi^{(n)(a)}_{\alpha\beta\ldots\nu} \, \xi^{(n')(b)}_{\alpha'\beta'\ldots\nu'} \tag{15}$$

where

$$\xi^{(n)(a)}_{\alpha\beta\ldots\nu} = (-1)^n (n!)^{-1} \sum_i e_i^{(a)} r_i^{2n+1} \frac{\partial}{\partial r_{i_\alpha}} \frac{\partial}{\partial r_{i_\beta}} \cdots \frac{\partial}{\partial r_{i_\nu}} r_i^{-1} \tag{16}$$

is the nth multipole moment operator of molecule a, and

$$\xi^{(0)(a)} = \sum_i e_i^{(a)} = q^{(a)} = \text{charge}, \tag{17}$$

$$\xi^{(1)(a)}_{\alpha} = \sum_i e_i^{(a)} r_{i\alpha} = \mu^{(a)}_{\alpha} = \text{dipole} \tag{18}$$

$$\xi^{(2)(a)}_{\alpha\beta} = \sum_i e^{(a)}_i \left(\tfrac{3}{2} r_{i_\alpha} r_{i_\beta} - \tfrac{1}{2} r_i^2 \delta_{\alpha\beta}\right) = \Theta^{(a)}_{\alpha\beta} = \text{quadrupole} \tag{19}$$

where r_{i_α} is the α component of the displacement of the ith charge of molecule a from an arbitrary origin (usually the centre of mass of a). The interaction tensor $\mathbf{T}$ in (15) is

$$T_{\alpha\beta\ldots\nu} = (4\pi\epsilon_0)^{-1} \, \nabla_\alpha \nabla_\beta \ldots \nabla_\nu R^{-1}, \tag{20}$$

so that

$$T = (4\pi\epsilon_0)^{-1} R^{-1}, \tag{21}$$

$$T_\alpha = -(4\pi\epsilon_0)^{-1} R_\alpha R^{-3}, \tag{22}$$

$$T_{\alpha\beta} = (4\pi\epsilon_0)^{-1} (3R_\alpha R_\beta - R^2 \delta_{\alpha\beta}) R^{-5}, \tag{23}$$

$$T_{\alpha\beta\gamma} = -3(4\pi\epsilon_0)^{-1} [5R_\alpha R_\beta R_\gamma - R^2 (R_\alpha \delta_{\beta\gamma} + R_\beta \delta_{\gamma\alpha} + R_\gamma \delta_{\alpha\beta})] R^{-7}. \tag{24}$$

From (13), (14) and (15), the longest-range interactions are

$$U^{(a)}_{\text{induction}} = -\tfrac{1}{2} \alpha^{(a)}_{\alpha\beta} F^{(a)}_\alpha F^{(a)}_\beta \tag{25}$$

$$U_{\text{dispersion}}(R^{-6}) = -T_{\alpha\beta} T_{\gamma\delta} \sum_{\substack{p_a \neq m_a \\ p_b \neq m_b}} \times \frac{\langle m_a m_b | \mu^{(a)}_\alpha \mu^{(b)}_\beta | p_a p_b \rangle \langle p_a p_b | \mu^{(a)}_\gamma \mu^{(b)}_\delta | m_a m_b \rangle}{\hbar(\omega_{p_a m_a} + \omega_{p_b m_b})} \tag{26}$$

where $\hbar\omega_{p_a m_a} = W_{p_a} - W_{m_a}$. In (25) $F^{(a)}_\alpha$, is the α component of the electric field at the origin in molecule a due to the unperturbed charge distribution of b:

$$F^{(a)}_\alpha = \sum_{n'=0}^{\infty} \frac{1}{1\cdot 3\cdot 5 \ldots (2n'-1)} T_{\alpha\alpha'\beta'\ldots\nu'} \langle m_b | \xi^{(n')(b)}_{\alpha'\beta'\ldots\nu'} | m_b \rangle$$

and $\alpha^{(a)}_{\alpha\beta}$ is the static polarisability tensor of molecule a:

$$\alpha_{\alpha\beta} = \alpha_{\beta\alpha} = \sum_{p \neq m} \frac{\langle m | \mu_\alpha | p \rangle \langle p | \mu_\beta | m \rangle + \langle m | \mu_\beta | p \rangle \langle p | \mu_\alpha | m \rangle}{W_p - W_m} \tag{27}$$

Unlike the electrostatic and induction energies, $U_{\text{dispersion}}$ does not separate simply into single-centre contributions. But this may be achieved (Dalgarno 1967) by using the identities (for $A > 0, B > 0$)

$$\frac{1}{A+B} = \frac{2}{\pi} \int_0^\infty \frac{AB}{(A^2 + \xi^2)(B^2 + \xi^2)} \, d\xi = \frac{2}{\pi} \int_0^\infty \frac{\xi^2}{(A^2 + \xi^2)(B^2 + \xi^2)} \, d\xi. \tag{28}$$

The polarisability at the frequency ω gives the linear dipole induced by an electric field of angular frequency ω:

$$\mu_\alpha(\omega, t) = \tilde{\alpha}_{\alpha\beta}(\omega) E_\beta = \tilde{\alpha}_{\alpha\beta}(\omega) E^{(0)}_\beta \exp(-i\omega t) \tag{29}$$

where

$$\tilde{\alpha}_{\alpha\beta}(\omega) = \sum_p \left[\frac{\langle m|\mu_\alpha|p\rangle\langle p|\mu_\beta|m\rangle}{\hbar(\omega_{pm}-\omega)} + \frac{\langle m|\mu_\beta|p\rangle\langle p|\mu_\alpha|m\rangle}{\hbar(\omega_{pm}+\omega)}\right] \tag{30}$$

$$= \sum_p \left[\frac{2\omega_{pm}\,\mathrm{Re}\,\{\langle m|\mu_\alpha|p\rangle\langle p|\mu_\beta|m\rangle\}}{\hbar(\omega_{pm}^2-\omega^2)} + \frac{2\mathrm{i}\omega\,\mathrm{Im}\,\{\langle m|\mu_\alpha|p\rangle\langle p|\mu_\beta|m\rangle\}}{\hbar(\omega_{pm}^2-\omega^2)}\right]$$

$$= \alpha_{\alpha\beta}(\omega) - \mathrm{i}\alpha'_{\alpha\beta}(\omega) = \alpha_{\beta\alpha}(\omega) + \mathrm{i}\alpha'_{\beta\alpha}(\omega). \tag{31}$$

Since $\dot{E}_\beta = -\mathrm{i}\omega E_\beta$, the antisymmetric polarisability $\alpha'_{\alpha\beta}$ gives the dipole induced by the time derivative $\dot{E}_\beta$. Since $\mu(\omega, t)$ and $\mathbf{E}$ are even under time reversal θ, while $\dot{\mathbf{E}}$ is odd, $\alpha_{\alpha\beta}(\omega)$ is even and $\alpha'_{\alpha\beta}(\omega)$ is odd under θ. Thus $\alpha'_{\alpha\beta}(\omega)$ exists only if the molecule is not symmetric under θ, as in an alkali metal atom with its unpaired electron or in any molecule in a magnetostatic field (Buckingham 1978).

On applying the identities (28) to (26), using $\alpha_{\alpha\beta}(\omega)$ in (31) at the imaginary frequency $\mathrm{i}\xi$ (notice that $\alpha_{\alpha\beta}(\mathrm{i}\xi)$ is well behaved mathematically and descends monotonically from the static polarisability at $\xi = 0$ to zero at $\xi \to \infty$):

$$U_{\text{dispersion}}\,(R^{-6}) = -\frac{\hbar}{2\pi} T_{\alpha\beta} T_{\gamma\delta} \int_0^\infty \{\alpha^{(\mathrm{a})}_{\alpha\gamma}(\mathrm{i}\xi)\alpha^{(\mathrm{b})}_{\beta\delta}(\mathrm{i}\xi) + \alpha'^{(\mathrm{a})}_{\alpha\gamma}(\mathrm{i}\xi)\alpha'^{(\mathrm{b})}_{\beta\delta}(\mathrm{i}\xi)\}\,\mathrm{d}\xi \tag{32}$$

which has achieved the desired separation and provides a rigorous formula for the longest-range dispersion energy in terms of an integral over all imaginary frequencies of the product of the polarisabilities of molecules a and b.

For an inert gas atom $\alpha'_{\alpha\gamma} = 0$ and $\alpha_{\alpha\gamma} = \alpha\delta_{\alpha\gamma}$ so that (32) reduces to

$$U_{\text{dispersion}}\,(R^{-6}) = -(4\pi\epsilon_0)^{-2}\frac{3\hbar}{\pi} R^{-6} \int_0^\infty \alpha^{(\mathrm{a})}(\mathrm{i}\xi)\alpha^{(\mathrm{b})}(\mathrm{i}\xi)\,\mathrm{d}\xi$$

$$= C_6 R^{-6} \tag{33}$$

Very reliable values of C_6 are now available (Dalgarno 1967, Tang *et al* 1976) and some are given in table 1.

Table 1. Values of the polarisability and of the coefficients C_6, C_8 and C_{10} in the expansion of the dispersion energy $U_{\text{dispersion}} = C_6 R^{-6} + C_8 R^{-8} + C_{10} R^{-10} + \ldots$ (Tang *et al* 1976). All values are in atomic units.

Atoms	$(4\pi\epsilon_0)^{-1}\alpha$	$-C_6$	$-C_8$	$-C_{10}$
H–H	4.50	6.499	124.4	3286
He–He	1.38	1.47	14.0	170
Ne–Ne	2.66	6.9	76	1100
Ar–Ar	11.1	67	1500	4×10^4
Kr–Kr	16.7	133	3200	9×10^4
Xe–Xe	27.3	260	6800	2×10^5
Na–Na	161	1510	1.1×10^4	11×10^6
Cs–Cs	409	7260	1.2×10^6	17×10^7

The contribution to $U_{\text{dispersion}}$ due to $\alpha'^{(a)}_{\alpha\gamma}\,\alpha'^{(b)}_{\beta\delta}$ is small and negative for a pair of alkali metal atoms having the same spin, and is positive when the spins are opposed. The corresponding small contribution to the interaction of rotating CH_4 molecules has been considered by Atkins (1980).

At intermediate separations R, it may be necessary to include higher multipole contributions to the dispersion energy. These can be handled similarly to the dipole–dipole term in R^{-6} (Dalgarno 1967), but in practice they are difficult to evaluate since spectroscopic intensities are not available for transitions involving higher multipoles.

For most molecules the dispersion of $\alpha_{\alpha\gamma}(\omega)$ is not accurately known so that (32) is not appropriate for evaluating $U_{\text{dispersion}}$. One may then use an approximate formula involving the *static* polarisabilities and some mean excitation energy u (usually taken to be the first ionisation energy) (London 1937, Buckingham 1967):

$$U_{\text{dispersion}} = -\frac{u_a u_b}{4(u_a + u_b)}\Big[T_{\alpha\beta}T_{\gamma\delta}\,\alpha^{(a)}_{\alpha\gamma}\alpha^{(b)}_{\beta\delta} + \tfrac{2}{3}\,T_{\alpha\beta}T_{\gamma\delta\epsilon}\,(\alpha^{(a)}_{\alpha\gamma}A^{(b)}_{\beta\delta\epsilon} - A^{(a)}_{\beta\delta\epsilon}\alpha^{(b)}_{\alpha\gamma})$$
$$+ \tfrac{1}{3}\,T_{\alpha\beta\gamma}\,T_{\delta\epsilon\zeta}\,(\alpha^{(a)}_{\alpha\delta}\,C^{(b)}_{\beta\gamma\epsilon\zeta} + C^{(a)}_{\beta\gamma\epsilon\zeta}\alpha^{(b)}_{\alpha\delta}) - \tfrac{2}{3}\,A^{(a)}_{\alpha\epsilon\zeta}A^{(b)}_{\delta\beta\gamma})$$
$$- \tfrac{2}{9}\,T_{\alpha\beta}\,T_{\gamma\delta\epsilon\zeta}\,A^{(a)}_{\alpha\gamma\delta}\,A^{(a)}_{\beta\epsilon\zeta} + \ldots\Big]. \qquad (34)$$

Here $A_{\beta\delta\epsilon}$ is a higher static polarisability that gives the dipole μ_β induced by an electric field gradient $\nabla_\delta E_\epsilon$ as well as the quadrupole $\Theta_{\delta\epsilon}$ induced by E_β, and $C_{\beta\gamma\epsilon\zeta}$ is the quadrupole polarisability giving $\Theta_{\beta\gamma}$ induced by the field gradient $\nabla_\epsilon E_\zeta$ (Buckingham 1967). For centrosymmetric molecules $A_{\beta\delta\epsilon}$ vanishes, but for a tetrahedral species such as CH_4 it provides an angle-dependent potential varying as R^{-7} and is the dominant cause of collision-induced rotational Raman scattering from compressed methane (Buckingham and Tabisz 1977). The approximate equation (34) may be obtained from (32) by relating the frequency-dependent polarisabilities to the static values by the simple formulae

$$\alpha_{\alpha\beta}(\omega) = \alpha_{\alpha\beta}\,[1 - (\hbar\omega/u)^2]^{-1} \qquad (35)$$

$$A_{\alpha\beta\gamma}(\omega) = A_{\alpha\beta\gamma}\,[1 - (\hbar\omega/u)^2]^{-1} \qquad (36)$$

$$C_{\alpha\beta\gamma\delta}(\omega) = C_{\alpha\beta\gamma\delta}\,[1 - (\hbar\omega/u)^2]^{-1} \qquad (37)$$

Crude estimates of α may be obtained by equating $(4\pi\epsilon_0)^{-1}\,\alpha$ to the cube of the molecular radius σ, and similarly $(4\pi\epsilon_0)^{-1}\,A \sim \sigma^4$ and $(4\pi\epsilon_0)^{-1}\,C \sim \sigma^5$.

4. Non-additivity

The earliest calculation of three-body dispersion forces was that of Axilrod and Teller (1943). The long-range dipole–dipole–dipole dispersion energy comes from third-order perturbation theory:

$$\Delta U_{abc} = U_{abc} - U_{ab} - U_{bc} - U_{ca} = D_{abc}\,(1 + 3\cos\theta_a\cos\theta_b\cos\theta_c)R_a^{-3}R_b^{-3}R_c^{-3} \qquad (38)$$

where R_a, R_b, R_c are the sides and $\theta_a, \theta_b, \theta_c$ the interior angles of the triangle formed by the atoms, and

$$D_{abc} = (4\pi\epsilon_0)^{-3} \frac{3\hbar}{\pi} \int_0^\infty \alpha_a(i\xi)\alpha_b(i\xi)\alpha_c(i\xi)\,d\xi. \tag{39}$$

For three H atoms $D_{HHH} = 21.64$ a.u. For three identical atoms at the vertices of an equilateral triangle

$$\frac{\Delta U_{abc}}{U_{ab} + U_{bc} + U_{ca}} = \frac{11DR^{-3}}{24C_6} \tag{40}$$

For atomic H at $R = 7$ a.u. $= 3.70 \times 10^{-10}$ m, $\Delta U/U = -0.00445$.
For argon at $R = 3.76 \times 10^{-10}$ m, $\Delta U/U = -0.010$.

For crystalline argon the dipole–dipole–dipole dispersion energy contributes –7.5% of the sublimation energy, which is approximately equal to the many-body contribution (which can be evaluated for Ar since the pair potential is accurately known) (Barker 1976). An explanation of the face-centred cubic structure of the inert gas solids Ne, Ar, Kr and Xe has been given in terms of a modified dispersion energy in the region of small overlap (Niebel and Venables 1974); the key fact is that the distortion of the atoms giving rise to the dispersion forces involves mixing excited d orbitals into the valence p orbitals, and the centrosymmetric fcc lattice is more favourably disposed to exploit this distortion than the hcp lattice.

5. Forces between macroscopic bodies

The force between two close bodies depends upon the long-range interaction energies between the constituent molecules. If R is large compared to the reduced wavelength $\lambda\!\!\!^- = c\omega^{-1}$ of the transitions responsible for the fluctuations in the charge density, the usual R^{-6} dispersion is 'retarded' and becomes

$$U = -(4\pi\epsilon_0)^{-2} \frac{23\hbar c}{4\pi} \frac{\alpha^{(a)}\alpha^{(b)}}{R^7}, R \gg \lambda\!\!\!^- \tag{41}$$

where $\alpha^{(a)}$ and $\alpha^{(b)}$ are *static* polarisabilities (Casimir and Polder 1948). The reason the static polarisabilities determine the retarded interaction, whereas dynamic polarisabilities determine C_6 (equation (33)), is that oscillations at frequencies $> c/R$ do not correlate effectively. At such large separations the forces are extremely weak and they are only significant for macroscopic bodies, possibly including colloids. These forces can be measured (Israelachvili and Tabor 1973) and calculated from bulk dielectric properties (Richmond 1975, Mahanty and Ninham 1976) in favourable cases.

The dispersion energy between a pair of $-CH_2-$ groups separated by 5×10^{-10} m is approximately -0.06×10^{-20} J. For two long parallel linear chains at a separation d, and each containing n CH_2 groups, the total dispersion energy varies as nd^{-5} and for $d = 5 \times 10^{-10}$ m is equal to $-0.3n \times 10^{-20}$ J $= -1.7n$ kJ mol^{-1} (Salem 1962). These forces provide a simple explanation of differences in the cohesive energy of *cis*-unsaturated fatty acids as compared to *trans*-unsaturated or saturated fatty acids (Salem 1962).

The heat of sublimation of crystalline CO_2 at 0 K is 27 kJ mol^{-1} and of this approximately 45% is due to the electrostatic quadrupole–quadrupole interactions and 55% to dispersion forces (Buckingham 1975).

6. Effects of dispersion-type interactions on molecular properties

The attractive dispersion force between two inert gas atoms must, by the Hellmann–Feynman theorem, be associated with a build-up of electron density in the region between the nuclei. In fact each atom loses its centre of symmetry and acquires a dipole moment proportional to R^{-7}; these dipoles cancel in the case of identical atoms but give a quadrupole moment proportional to R^{-6}. If one were to evaluate the attractive force $-6C_6R^{-7}$ through this distorted charge distribution ρ, it would be necessary to know the second-order changes in ρ up to those varying as R^{-7} (Hirschfelder and Meath 1967). However, to obtain the force through the perturbed energy C_6R^{-6} it is only necessary to know the first-order change in ρ varying as R^{-3}.

Optical, electric and magnetic properties of atoms are affected by the collision-induced change in ρ due to the dispersion interaction. The mean polarisability of a pair of atoms changes as R^{-6} at large R, and for light atoms the dispersion-type contribution is larger than that due to classical dipolar interaction (Buckingham and Clarke 1978). The dipole of a pair of dissimilar atoms varying as R^{-7} is entirely due to dispersion-type interaction; it can approximately be evaluated from the higher static polarisability $B_{\alpha\beta\gamma\delta}$ giving the effect of a field gradient $\nabla_\gamma E_\delta$ on the polarisability $\alpha_{\alpha\beta}$ (Hunt 1980) and, in a rigorous formula, from the frequency-dependent $B_{\alpha\beta\gamma\delta}(\omega)$ and $\alpha_{\alpha\beta}(\omega)$ at imaginary frequencies (Galatry and Gharbi 1980).

References

Atkins P W 1980 *Chem. Phys. Lett.* **74** 358–361
Axilrod B M and Teller E 1943 *J. Chem. Phys.* **11** 299–300
Barker J A 1976 *Rare Gas Solids* ed. M L Klein and J A Venables (London: Academic Press) ch.4
Buckingham A D 1967 *Adv. Chem. Phys.* **12** 107–142
— 1975 *Phil. Trans. R. Soc.* **B 272** 5–11
— 1978 *Intermolecular Interactions: From Diatomics to Biopolymers* ed. B Pullman (Chichester, Sussex: Wiley) ch.1
Buckingham A D and Clarke K L 1978 *Chem. Phys. Lett.* **57** 321–325
Buckingham A D and Tabisz G C 1977 *Opt. Lett.* **1** 220–222
Casimir H B G and Polder D 1948 *Phys. Rev.* **73** 360–372
Dalgarno A 1967 *Adv. Chem. Phys.* **12** 143–166
Debye P 1921 *Phys. Z.* **22** 302–308
Galatry and Gharbi T 1980 *Chem. Phys. Lett.* **75** 427–433
Hirschfelder J O and Meath W J 1967 *Adv. Chem. Phys.* **12** 3–106
Hunt K L C 1980 *Chem. Phys. Lett.* **70** 336–342
Israelachvili J N and Tabor D 1973 *Prog. Surf. Membrane Sci.* **7** 1–55
Langbein D 1974 *Theory of Van der Waals Attraction* (Springer Tracts in Modern Physics **72**)
Liftshitz E M 1956 *Sov. Phys.–JETP* **2** 73–83
London F 1937 *Trans. Faraday Soc.* **33** 8–26
Longuet-Higgins H C 1965 *Discuss. Faraday Soc.* **40** 7–18
McLachlan A D 1965 *Discuss. Faraday Soc.* **40** 239–245
Mahanty J and Ninham B W 1976 *Dispersion Forces* (London: Academic Press)
Niebel K F and Venables J A 1974 *Proc. R. Soc.* **A 336** 365–377
Richmond P 1975 *Colloid Science* ed. D H Everett (Chem. Soc. Spec. Period. Rep. **2**) pp130–172
Salem L 1962 *Can. J. Biochem. Physiol.* **40** 1287–1298
Tang K T, Norbeck J M and Certain P R 1976 *J. Chem. Phys.* **64** 3063–3074

Inst. Phys. Conf. Ser. No. 58
Invited paper presented at Physics of Dielectric Solids, 8–11 September 1980, Canterbury

Contact charging of dielectrics

A C Rose-Innes
Joint Laboratory of Physics and Electrical Engineering, University of Manchester Institute of Science and Technology, Manchester M60 1QD, UK

Abstract. Virtually all substances acquire a charge when touched by a different substance. This 'contact charging' is of particular importance in insulators because charge transfer to them does not leak away and may build up to high values. Contact charging of insulators is of great significance in industry, where it may be either a nuisance (e.g. marking of photographic film), or useful (e.g. the Xerox photocopying process); but in spite of its industrial importance and of the fact that it is the oldest known electrical phenomenon, why contact charging occurs is little understood.

Recently there has been an increase in research into the fundamental processes of contact charging and, as a result of advances in experimental techniques, a body of reliable experimental data is now being built up. This review concentrates on the contact charging of insulators by metals, reviewing recent experimental work aimed at revealing the nature of the states in the insulators which accept electrons from, or give electrons to, the contacting metal.

1. Introduction

It is a common observation that if an insulator is rubbed by another substance it becomes electrically charged. Surprisingly, in spite of the fact that this 'contact charging' is the oldest studied electrical phenomenon, little is known about why charge transfers between the two substances, and it remains a largely unsolved problem in solid state physics. Indeed, at first sight contact charging seems a rather improbable phenomenon, because two oppositely charged bodies have a higher free energy than if they were uncharged. There is no thermodynamic law which prevents this situation (work must be performed to separate the bodies which have exchanged charge); nevertheless we may be a little surprised that it should happen.

Before considering this problem in any detail let us note two simple facts. First, *rubbing* is not necessary for contact charging. Quite large charges (up to about 10^{-4} C per square metre of contact) may transfer between two substances which have been touched together without any tangential component of relative motion. In this review I shall consider only contact charging from such 'normal contacts', because rubbing introduces ill defined complications, such as local heating and surface damage. The second fact is that contact charging is not confined to insulators. Nevertheless, it is particularly noticable on insulators; this is because charge transferred to them does not leak away. If two *metals* are touched together the charge transferred from one to the other can be as great as or even greater than that transferred between two insulators but, under most circumstances, the high conductivity of the metals allows the charge to leak away very quickly from the region of contact.

0305-2346/81/0058-0122 $01.50

In trying to understand the reason for contact charging we are naturally interested in where the transferred electrons go†, that is to say, in the nature, energy and position of the sites in the insulator which accept electrons from, or donate electrons to, the contacting metal. Discussion of the nature of the electron sites must be based on experimental evidence and so I now consider some aspects of experimental techniques for measurement of contact charging.

2. Experimental

As in all fields of experimental investigation, there are a number of pitfalls waiting for the unwary. Unfortunately, failure to recognise some of these has, in the past, led to the publication of incorrect results, and this is perhaps why contact charging has the reputation of being a topic in which it is impossible to get reproducible results.

The task is to measure the charge on a specimen which has been touched by another specimen. In the case of insulators, contact charge is on, or close to, the surface and is normally measured by *induction* (Figure 1*a*). The principle is simple: suppose we have a

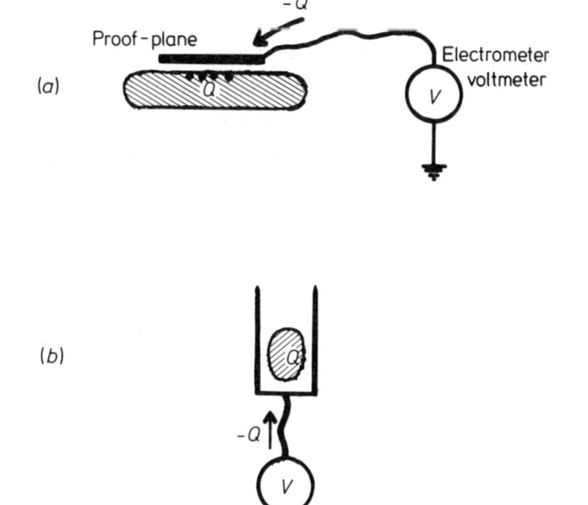

Figure 1. Measurement of surface charge: (*a*) by proof plane, (*b*) by Faraday pail.

small area of surface with a charge Q on it; we bring up to the surface a metal plate (called a 'proof-plane') which is connected to a very-high-resistance voltmeter (electrometer). The potential of the proof-plane is raised by the field from the charge on the insulator to some value V which is indicated on the voltmeter. The contact charge is then given by $Q = C'V$ where C' is the appropriate combination of the capacitances of the charged surfaces to earth and to the proof-plane, and of the capacitance of the metal plate to

†In this review I make the assumption that contact charging is the result of *electrons* transferring from one body to the other. Contact charging could, however, occur by transfer of surface ions or of charged material, and a few workers have suggested these as the basic cause. Though ion or material transfer may be significant in a few special cases, the evidence strongly suggests that contact charging is normally by electron transfer. There is not room in this paper to review the evidence on this point but it will be discussed in a forthcoming review (Lowell and Rose-Innes 1980).

earth (including the input capacitance of the voltmeter). This effective capacity may not, however, be easy to calculate because the polarisation of the insulator may not be simply related to the electric field in it (i.e. insulators do not always have a unique permittivity). Furthermore, the insulator may be polarised for reasons that have nothing to do with contact charge. Polymers which have been exposed to an electric field may acquire a polarisation which lasts for a very long time. Now any dielectric polarisation is equivalent to the presence of charges of opposite sign on opposite faces of the specimen and this *polarisation charge* on any face will induce charge on a proof-plane brought up to that face. So the voltage indicated by the electrometer could be partly due to a polarisation unconnected with contact charge. If we want to be certain that we are measuring only the charge transferred *to* a body and not any charge separation within it we must insert the whole specimen into a Faraday pail connected to a voltmeter, as shown in figure 1(*b*).

Another experimental point concerns the amount of charge. The amount of contact charge transferred to a surface can be quite large even if there is no rubbing. Typically one needs to measure charge of about 10^{-10} C, which is quite easy with modern electrometers, which can measure about 10^{-13} C. Though the relatively large contact charge should make measurement easy it may, unfortunately, also make the measurement worthless! The trouble is that the initial contact charge density on the insulator surface can be so high that is dissipates itself by sparking or corona discharge into the surrounding air before it can be measured. The breakdown field of air limits any charge density remaining on the surface to less than about 10^{-4} C m^{-2}, though the density of the original charge transferred onto it may have been much more than this. Figure 2(*b*) shows how irreproducible contact charge measurement can be if carried out in the atmosphere. This loss of charge by gaseous discharge can be eliminated if the experiments are carried out in a vacuum. A pressure of less than about 10^{-4} mb virtually eliminates any loss of charge by gaseous discharge (figure 2*a*). For this reason, if contact charge measurements are to be reliable they should be carried out in a vacuum chamber.

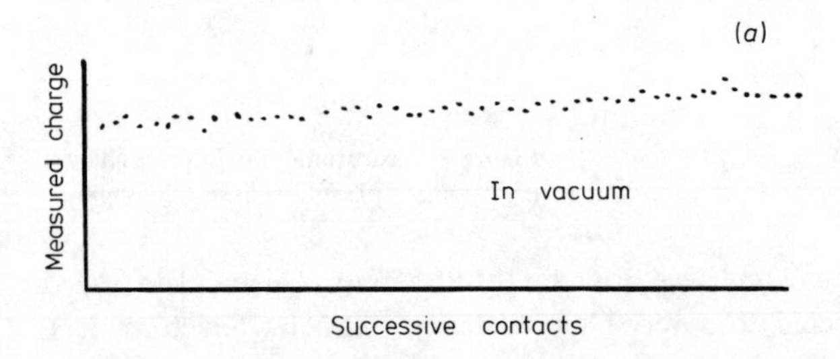

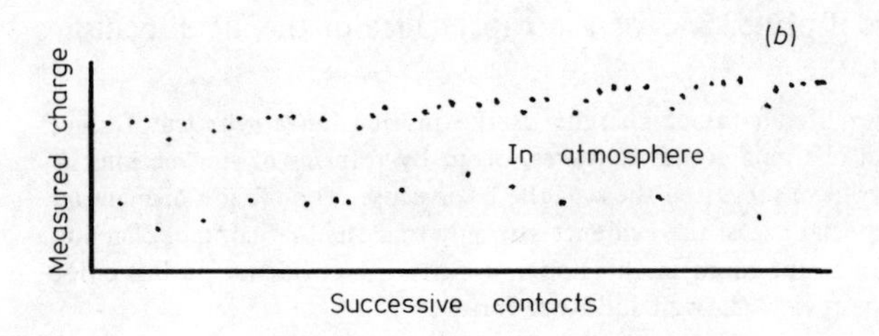

Figure 2. Charge measured on insulator for repeated contact by metal ball to same spot: (*a*) in vacuum; (*b*) in atmosphere.

3. Contact charging of dielectrics

The most spectacular examples of contact electrification occur when one insulator is touched by another insulator. However, it is difficult to study these cases because at present we know too little about the electron states in insulators, especially at their surface. On the other hand, the electron states in metals are rather simple and well understood and, as we shall see, the contact charging of one metal by another behaves in just the way we would expect from our understanding of their electronic structure. For this reason most work on contact electrification of insulators is at present carried out by touching them with metals; interpretation of results is obviously easier if the behaviour of at least one of the two contacting materials is understood.

Let us begin by considering the contact charging of one *metal* by another. In metals there is a band of allowed electron states which is filled up to the Fermi energy E_F (figure 3*a*). The workfunction ϕ of the metal is the amount of energy by which its Fermi level lies below the vacuum level and, in general, two metals, A and B, will have different workfunctions, ϕ_A and ϕ_B. When the two metals touch, electrons flow from the metal with the higher Fermi energy into that with the lower Fermi energy (figure 3*a*). This electron

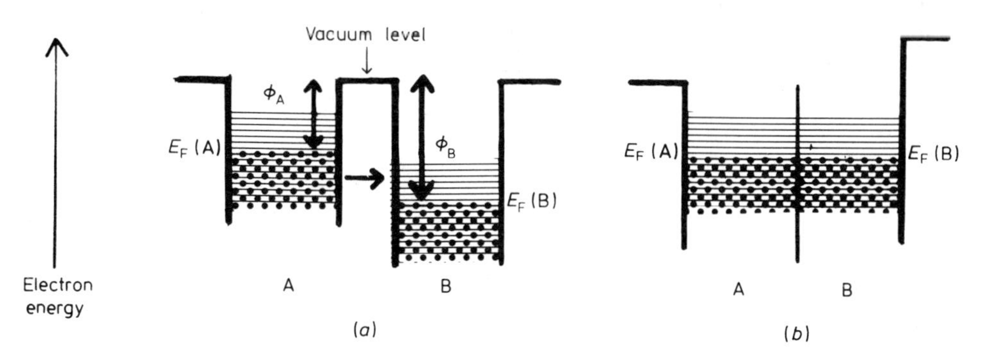

Figure 3. Two metals, A and B: (*a*) before contact; (*b*) after contact.

transfer raises the electron potential of B relative to A and ceases when the Fermi energy of B has risen to equal that of A (figure 3*b*). According to this model the charge transfer should be proportional to the difference between the workfunctions of the two metals,

$$Q \propto (\phi_B - \phi_A) \tag{1}$$

and, indeed, this is what is found by experiment. So the contact charging of metals behaves just as we should expect from our knowledge of their rather simple structure.

3.1. Charging of insulators by metals

Some insulators acquire negative charge when touched by a metal, others acquire positive charge. However, to save tedious repetition, I shall in this review consider only negative charging, i.e. electrons transferring *to* the insulator from the metal. For every situation discussed there is an analogous situation in which the insulator can lose electrons to the metal; for example, when we consider empty acceptor states in the insulator lying below the Fermi level of the contacting metal, we must remember that there is the opposite situation in which the insulator contains full donor states lying above the metal's Fermi energy.

The term 'insulator' covers a very wide range of materials with great differences in structure. Polyethylene, glass and sapphire are all insulators but, apart from being bad conductors of electricity, seem to have little in common. However, the fact that they are all bad conductors of electricity tells us that they do have one essential common feature. There is a gap in the spectrum of allowed electron energies (figure 4), a valence 'band' of

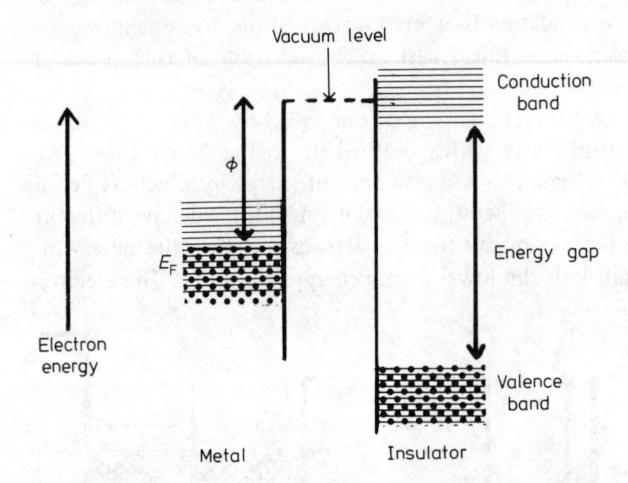

Figure 4. Metal and insulator.

states fully occupied by electrons lying several electron volts below a conduction 'band' of unoccupied states. The nature of the 'bands' depends on the kind of insulator; in a crystalline, strongly bonded material, such as sapphire, the bands will consist of nonlocalised states extending throughout the specimen, whereas in a disordered weakly bonded material, such as a polymer, the band will consist of states that are partly localised. However, for our purposes the difference is not important, and I shall refer to both kinds simply as 'bands'. Now there is evidence that in most insulators, the bottom of the conduction band lies close (within an eV or so) to the vacuum level (see, for example, Mort 1980, Schwentner *et al* 1975, Williams 1965) and this leads to the interesting conclusion that the contact charging of such insulators by metals is impossible! Consider a metal and an insulator (figure 4). The work functions, ϕ, of metals have values between about 3 and 6 eV, and this means that the Fermi level of a metal lies in the energy gap of the insulator, well below the bottom of the insulator's conduction band. Therefore electrons should not be able to pass from the metal into the insulator; they cannot go into the conduction band because this is full and, at room temperature T_R, there is far too little thermal energy to excite them up into the empty conduction band ($kT_R \sim 3 \times 10^{-2}$ eV). Nevertheless, it is an experimental fact that virtually all insulators *do* charge when touched by metals. The answer to this apparent paradox must be that in real insulators there are electron-accepting sites with energies inside the energy gap. Such sites could be impurities, imperfections or surface states. Electrons from a contacting metal must reach the acceptor sites in the insulator by tunnelling, so only true surface states or those bulk impurity sites which happen by chance to be very close to the surface (say not deeper than 0.3 nm) will take up electrons.

We can consider two cases. In one case the density of acceptors in the insulator is so large that, when they are filled by electrons from a contacting metal, the resulting electric field between it and the insulator is significant. In this case electron transfer will cease when the resulting potential step across the interface has increased to such a value that it has raised the energies of the acceptor sites above the Fermi energy of the metal. A rough calculation shows that for this situation to occur the acceptor concentration must be such as to give a contact charge density of about $10^{-2}\,\mathrm{C\,m^{-2}}$. At the other extreme, the density of sites in the insulator may be so low that, even if they have all taken up electrons, the resulting electric field across the interface is so small that there is a negligible increase in their energy. In this case the magnitude of contact charge will be limited by the density of states in the insulator; all states accessible by tunnelling are filled on contact. Experiments to determine absolute contact charge density are not easy because it is difficult to measure the area of actual contact; however, those experiments which have been done suggest that, on polymers, contact charge densities are not more than about $10^{-4}\,\mathrm{C\,m^{-2}}$. These results suggest, therefore, that contact charging is limited by filling up the available acceptor sites in the insulator. Incidentally, it has been found in experiments on insulators which have been deliberately doped with electron-accepting or -donating impurities (Lowell 1979, Cottrell *et al* 1979) that the number of electrons taken up from a contacting metal is several orders of magnitude less than the number of acceptor impurities which lie within tunnelling distance of the surface. We do not know why so few of the sites which should be able to accept electrons do in fact do so.

What else can experiment tell us about the sites in an insulator that take up electrons in contact electrification? One kind of experiment consists in touching one insulator with a number of different metals. Typical experimental results are presented in figure 5, which shows that the contact charge density deposited on the insulator varies smoothly with the workfunction† of the contacting metals and that the amount of negative charge deposited

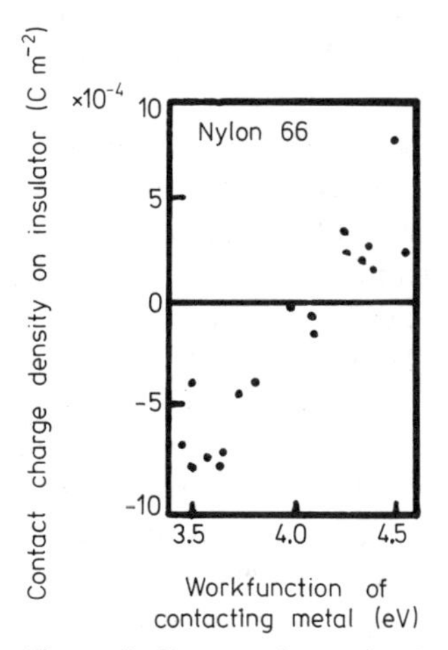

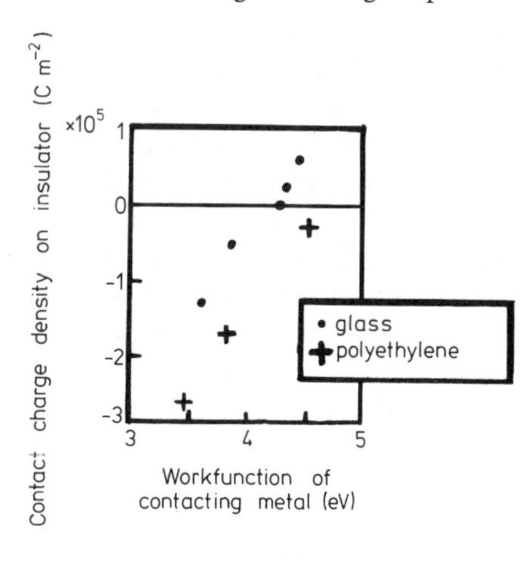

Figure 5. Contact charge density on insulators touched by metals of different workfunctions (after Davies 1967, 1969).

†It is important in these experiments to measure the workfunctions of the actual metal specimens used and not to rely on published values, because the workfunction of a metal may depend strongly on small variations in the condition of its surface.

increases as the workfunction of the contacting metals gets smaller (i.e. higher Fermi energies).

Any variation of contact charge with the workfunction of the metals should give information on the energy of the acceptor sites because only those insulator levels which are below the metal's Fermi level can acquire electrons. This approach was pioneered by Arridge (1967) and by Davies (1967). Again we can consider the limiting cases. At one extreme, all the electron-acceptor sites in the insulator have the same energy, so they all lie above or below the Fermi level of any one contacting metal. At the other extreme, the energies of the electron-acceptors may be spread over a broad band so that for each contacting metal a different number lie below its Fermi level.

A theoretical prediction about the electron energy levels lying within the energy gap of certain polymers has been made by Duke and Fabish (1976). Polymers (a popular kind of insulator on which to carry out contact charging experiments) can be divided into two classes: those (such as polystyrene) which have side groups attached to the carbon backbone and those (such as polyethylene) which do not. A distribution of electron energy states within the energy gap has been predicted by Duke and Fabish for those polymers which have side groups attached to their carbon backbone (e.g. the benzene ring in polystyrene or the –OH group in polyvinyl alcohol). Electrons can become attached to or removed from these side groups, and the corresponding acceptor or donor levels form a band about 4 eV below the vacuum level and about 1 eV wide. Note that, according to this model, the states within the energy gap are not associated with impurities or defects but are an inherent property of the polymer itself. It is interesting to note in connection with Duke and Fabish's model of inherent energy states associated with side groups that, whereas the contact charging of polymers with side groups depends systematically on the workfunction of the contacting metal, experiments by Lowell (1976) show that the contact charging of polyethylene and polytetrafluoroethylene, which do not have side groups, is independent of the metal workfunction. However, Lowell's results for these two polymers are not in agreement with some earlier published results (e.g. Davies 1967) and this important point needs experimental verification.

It can be seen from figure 5 that an insulator may charge negatively when touched by a metal of small workfunction (high Fermi energy) but positively when touched by a metal of larger workfunction (lower Fermi energy).This implies that such an insulator contains both empty acceptor states and, below these, filled donor states. The energies of filled donor sites which are on or very close to the surface can, in principle, be measured by photoemission. Photoemission experiments on good insulators are not easy, however, because the low conductivity prevents the establishment of a steady-state photocurrent. Nevertheless Dr B Hamilton of my laboratory has recently made such measurements on polyvinyl acetate, a polymer which charges positively (i.e. loses electrons) when contacted by gold, cadmium or aluminium. He obtained photoemission of electrons with radiation whose photon energy was greater than 3.1 eV, which shows there are electron-containing states at this energy below the vacuum level. However, polyethylene which charges negatively (acquires electrons) when contacted by metals showed no photoemission for irradiation by photons with energies up to about 5 eV. Murata and colleagues (Murata 1979, Murata *et al* 1979) have also used optical irradiation to investigate the energies of electron sites responsible for contact charging. From their experiments they draw the conclusion that in some polymers there is a broad band of energy levels spanning the

Fermi energies of typical metals and that within any small range of energies some of these levels are occupied by electrons while others are not, i.e., there are electrons in metastable excited states. In one experiment (Kittaka and Murata 1979) polyethylene and polypropylene, which normally acquire a negative charge when contacted by gold, were irradiated with ultraviolet light before being contacted by the metal. These irradiated samples were then found to charge positively when contacted. Kittaka and Murata's interpretation of this result was that the ultraviolet light had excited electrons to states in the insulator lying above the Fermi energy of the contacting metal so that, on contact, these electrons transferred *to* the metal. There seems, however, to be a difficulty with this explanation: the electrons photoexcited to high energy in the insulator would leave behind an equal number of empty states which would be *below* the Fermi level of the contacting metal. On contact, therefore, electrons should also transfer *from* the metal into these empty states, and, consequently, there should be no net effect of the photoexcitation on the contact charging.

3.2. Absence of equilibrium: multiple contacts

Many theories of contact electrification are based on the assumption that during contact between a metal and an insulator the electrons in the insulator come into thermodynamic equilibrium with those in the metal. For example, Davies (1967) explained his result that the contact charging of one insulator varies linearly with the workfunction of the contacting metal by assigning to every insulator a 'Fermi level', the charge transfer being such as to bring this into coincidence with the Fermi level of the contacting metal. However, insulators are, almost by definition, materials in which the electrons, if disturbed, do not return to their equilibrium distribution in a reasonable time. Indeed, we have already seen that there is evidence that in polymers electrons may be sitting in excited metastable states. It seems therefore that we should not attempt to explain the contact charging of insulators in terms of their Fermi levels coming into equilibrium with that of a contacting substance.

An experimental result which has an important bearing on the questions as to whether local equilibrium is achieved during contact is the repeated contact of a metal to the same spot on an insulator (Lowell 1976, Fabish *et al* 1976). Suppose, for example, that a metal ball touches the plane surface of an insulating plate (this particular combination is often used); when the metal and insulator are separated a charge, Q_1 say, is found to have transferred to the insulator. If now the two are contacted again so that the metal ball touches the *same* spot on the insulator it is found that on separation the contact charge has increased, i.e. additional charge Q_2 has transferred on the second contact. Yet more charge transfer occurs on subsequent contacts (figure 6). The rate of increase of the charge decreases as contacts are repeated, and on a linear plot (figure 6*a*) the charge appears to saturate after a few hundred contacts. However, a plot of the accumulated charge against the *logarithm* of the number of contacts (figure 6*b*) shows that this is not so. In fact the charge has been found to be still increasing after many thousands of contacts. This build-up of charge with repeated contact is a general phenomenon; it has been observed on many insulators of quite different types; for example, polyethylene, polytetrafluoroethylene, diamond, corundum and anthracene. So far it has not been possible to find a general explanation of this effect. One explanation, which at first sight

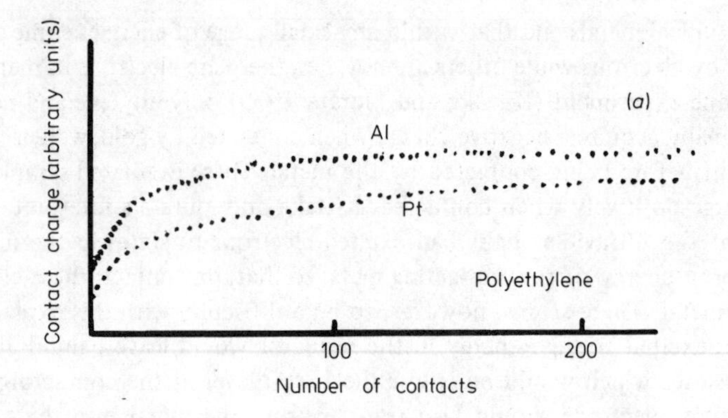

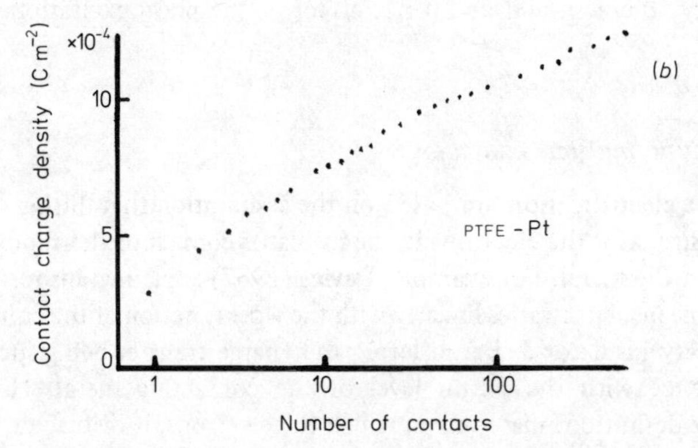

Figure 6. Build-up of charge with successive contacts. (*a*) Polyethylene; linear contact scale. (*b*) Polytetrafluorethylene, logarithmic contact scale.

seems rather obvious, would be that, during contact, charge crosses rather slowly from the metal to the insulator so that the increasing charge is just due to an increase in the total time of contact. Experiments (Lowell 1976, Homewood, private communication) show, however, that this is not the case: the magnitude of the accumulated charge, Q, does not depend on how long each contact lasts nor on the time between contacts, but only on the *number* of contacts. In other words, to increase the charge we must repeat the contacts, not merely prolong them. An alternative 'obvious' explanation is that, on each contact, the two contactors do not touch in quite the same place, so that additional charge is deposited on previously uncontacted areas. Again, however, experiments show that this is not the case; in carefully designed apparatus any increase in contacted area on repeated contacts is much too small to account for the large (up to 10-fold) increase in the charge after many contacts.

The fact that the accumulated charge depends on the number of contacts and not on the total time of contact suggests that the very action of contacting (or separating) contributes to the transfer of charge. For example, the pressure of contact might create some local mechanical distortion. It is interesting that repeated contact by liquid mercury to polyethylene and sapphire (Homewood, private communication) does not result in a build-up of charge after about the second or third contact, whereas charge builds up in the usual way when contacts are made by solid metals.

A general explanation of the build-up of charge on insulators by repeated contacts to a metal has not yet been found. Possible mechanisms which might be involved include: (i) a weak surface conduction which, when the metal is withdrawn and the transferred charge is no longer attracted to its image charge in the metal, allows the charge to spread out from the points of contact; (ii) formation of electron traps by the stress of contact, e.g. breaking of bonds in the insulator; (iii) 'stirring' of the insulator surface by plastic deformation at the contact so that filled electron traps are driven into the insulator and unfilled traps rise to the surface (Lowell 1976). Unfortunately in this review we do not have the space to review all possible explanations of the effect or the evidence for or against each of them (these are discussed in detail in a forthcoming article by Lowell and Rose-Innes (1980); at present no explanation seems to account adequately for all the aspects of the phenomenon. When seeking an explanation for the build-up of charge, we must be careful not to make an unwarranted assumption: the phenomenon is observed on a wide variety of insulators, but it does not necessarily follow that the explanation is the same for different kinds of insulator. Several quite different mechanisms could give rise to a build-up with the general form shown in figure 6. For example, there is evidence (Homewood, unpublished) that the build-up of charge on corundum is related to a weak electrical conductivity, whereas the build-up of charge on polyethylene is connected with deformation at the contact.

Even though we lack an explanation, this build-up of charge with repeated contact does tell us something very definite and useful. Because the charge after any contact is less than the charge after subsequent contacts, the material at the contact cannot have come into equilibrium during the contact (if equilibrium were achieved, the charge would be the same after all contacts). Consequently theories which assume that the electrons in the contacting bodies come into thermodynamic equilibrium (such as those in which the insulator is given a Fermi level which comes into equilibrium with the contacting metal) must be treated with suspicion.

4. Conclusion

Contact charging of insulators is of enormous importance in industry, both because it can be a nuisance and because it can be useful, but little is understood about why it occurs. In the past this topic has had the reputation of being one in which it is very difficult to get reproducible experimental results, but improvements in techniques, especially the realisation that measurement should be carried out in vacuum, have now made it possible to get reliable data.

The contact charging of insulators appears to be due to the acceptance or donation of electrons by sites whose energy lies deep in the energy gap. These could be imperfections and impurities, surface states or, in the case of polymers, states associated with side groups. Unfortunately, little is known about these electron states in most insulators and for this reason contact charging of insulators is often studied by touching them with metals, solids whose electronic structure and contact behaviour are rather well understood.

Unfortunately, a considerable proportion of experiments on contact charging of insulators have been performed on polymers, presumably because of their practical importance and because they acquire rather large charges. However, it is difficult to interpret the results because relatively little is known about the electronic structure of most polymers. An understanding of the general phenomenon of contact charging will probably be achieved quicker if more experiments are conducted on better characterised insulators such as inorganic crystals (e.g. corundum, alkali halides). This is an approach which we are adopting in my laboratory.

The build-up of charge by repeated contacts to the same spot on the insulator is an important but unexplained phenomenon. Though it seems to occur on all insulators it is not clear whether there is a general explanation or whether different processes are at work in different kinds of insulator. The build-up does show, however, that equilibrium is not established on a single contact and theories of contact charging must not be based on the assumption that equilibrium has been achieved.

In the space of this short paper I have not been able to discuss a number of aspects of contact charging†. For example, almost any theory of the magnitude of the contact charge will be concerned with the charge transferred *per unit area of contact.* Now it is not difficult to measure the total charge transferred, but it is usually very difficult to determine the area of actual contact between two bodies. Because the surfaces of all solids are rough on a fine scale, the actual area of contact will be less than the apparent area of contact. This difficulty in determining the area of contact is a serious hindrance in comparing experimental results with each other and with theory.

In summary, a body of reasonably reliable experimental data on the contact charging of insulators is now beginning to accumulate but the interpretation of this is often far from clear. Contact charging still offers a challenging field for the ingenious experimenter and the imaginative theoretician.

Acknowledgment

I am very grateful to my colleague, Dr John Lowell, for help in the preparation of this article.

References

Arridge R C G 1967 *Br. J. Appl. Phys.* **18** 1311
Cottrell G A, Reed C and Rose-Innes A C 1979 *Electrostatics 1979* (Inst. Phys. Conf. Ser. 48) p249
Davies D K 1967 *Static Electrification* (Inst. Phys. Conf. Ser. 4) p29
Davies D K 1969 *Br. J. Appl. Phys.* **2** 1533
Duke C B and Fabish T J 1976 *Phys. Rev. Lett.* **37** 1075
Fabish T J, Saltsburg H M and Hair M L 1976 *J. Appl. Phys.* **47** 940
Kittaka S and Murata Y 1979 *Jap. J. Appl. Phys.* **18** 515
Lowell J 1976 *J. Phys. D: Appl. Phys.* **10** 1571
— 1979 *J. Phys. D: Appl. Phys.* **12** 2217
Lowell J and Rose-Innes A C 1980 *Adv. Phys.* **29** 947
Mort J 1980 *Adv. Phys.* **29** 367
Murata Y 1979 *Jap. J. Appl. Phys.* **18** 1
Murata Y, Hodoshima T and Kittaka S 1979 *Jap. J. Appl. Phys.* **18** 2215
Schwentner N, Himpsel F-J, Skibowski M, Steinmann W and Kock E E 1975 *Phys. Rev. Lett.* **34** 528
Williams R 1965 *Phys. Rev.* **140** A569

†Many of these aspects will be covered in a forthcoming review (Lowell and Rose-Innes 1980).

Inst. Phys. Conf. Ser. No. 58
Invited paper presented at Physics of Dielectric Solids, 8–11 September 1980, Canterbury

Bulk effects in charge trapping

G M Sessler
Technische Hochschule Darmstadt, Darmstadt, W Germany

Abstract. A number of dielectrics are capable of charge-trapping effects. The most permanent charge-storage phenomena are encountered in polymer materials, some of which have extremely low conductivities and deep trapping centres. Many aspects of charge trapping and its slow change with time have been investigated recently utilising a variety of experimental and analytical methods. A distinction between surface and bulk traps, and a measurement of the activation energies of these levels was achieved by thermally stimulated current, thermal pulse, and electron-beam techniques. The radiation-induced conductivity and its time dependence as well as radiation-generated polarisation effects were investigated with electron-beam and thermally stimulated current methods. Some of these effects were also analysed theoretically. Finally, electron-beam probing was used very recently to determine the actual charge distribution in polymers.

1. Introduction

Charge trapping in highly insulating dielectrics has been of considerable interest recently. The growth of activities in this field is partially due to the usefulness of charged materials, commonly called electrets, in an increasing number of applications. Quite apart from this, research on charge trapping in dielectrics has opened up new approaches to the investigation of solid state properties of highly insulating materials. This was aided by the development of a number of new methods specifically designed for the investigation of charged dielectrics, such as thermally stimulated current (TSC) techniques and electron-beam probing.

The present paper is primarily concerned with bulk trapping phenomena in a class of dielectrics which is capable of relatively permanent charge storage, namely polymers. Of specific interest are thin polymer films of thickness between 6 and 25 μm which have found a variety of applications.

In the following, we review first the controlled charging of polymers. Then, recent investigations of geometrical and energetic trap depths, and of trap densities, together with the implications of these studies with respect to charge decay, are discussed. This is followed by a review of new analytical and experimental data on the effect of radiation on charge trapping and conductivity. Finally, a powerful method for the probing of the distribution of trapped charges by means of an electron beam is introduced and a result obtained with this method is depicted.

2. Charging of polymers

The investigation of charge-trapping effects in dielectrics depends to a large degree on the availability of reliable charging methods. A number of techniques, based on corona

0305-2346/81/0058-0133 $01.50

discharges, electron beams, liquid contacts, application of electric fields at elevated temperatures, etc, have been described (see, e.g., Sessler 1980). In the present context, the corona and electron-beam methods are of particular interest because they afford a great deal of control. The set-up for these methods is shown in figure 1.

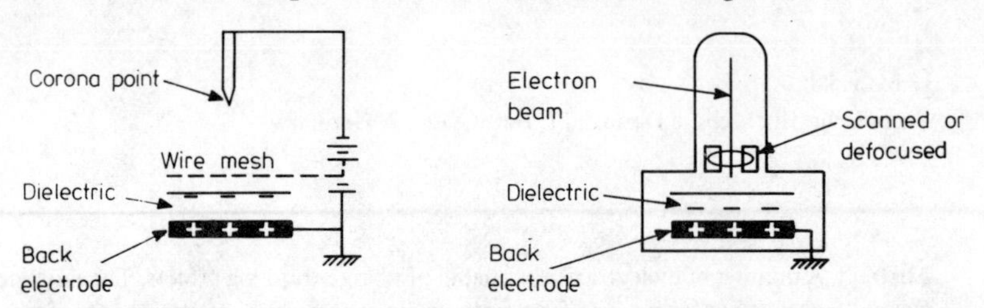

Figure 1. Schematic representation of the corona charging method (left) and the electron-beam charging method (right).

In the corona method, a one-sided metallised dielectric is exposed to a corona discharge generated by application of a voltage of a few kV to a needle electrode close to the nonmetallised surface of the sample. By means of this process, charges are deposited onto the exposed surface. A number of recent investigations (see, e.g., Moreno and Gross 1976) have shown that the penetration of these charges is very different for different polymer materials. Without further measures, the distribution of the charges in the plane of the sample exhibits maximum density in the centre and smaller (or zero) density on the edge. The distribution can be made laterally uniform by interposing a biased grid between the point electrode and the sample, as shown in figure 1 (Ieda *et al* 1967, Moreno and Gross 1976). Even in this case, however, the uniformity develops only gradually with time as the surface potential of the dielectric reaches the grid potential (Gerhard 1981). An example of this build-up is shown in figure 2. The spatial fluctuations initially seen are due

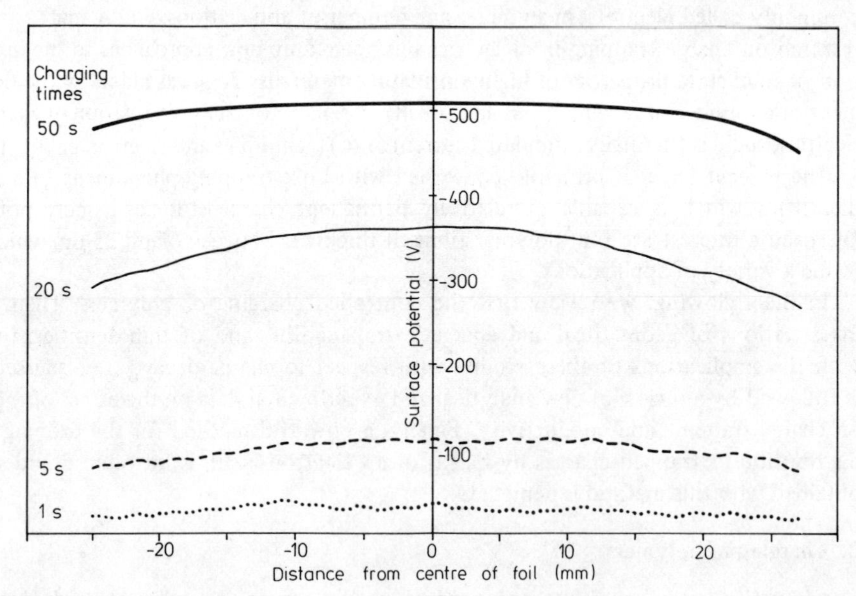

Figure 2. Charge build-up due to corona application to 25 μm fluoroethylenepropylene (FEP) films. Corona voltage, –6 kV; grid voltage, –0.5 kV (Gerhard 1981).

to the grid wires. Under the conditions of this experiment, most of the inhomogeneities disappear after about 50 s.

The electron-beam method offers a certain control over the charge penetration by means of the electron energy. Also, the lateral distribution of the charge is very uniform if the beam has a uniform density over its cross section (Sessler and West 1974). These features have made this method a preferred tool for the charging of dielectrics for research purposes and for commercial applications.

3. Trap levels in polymers and thermally stimulated charge decay

Most polymer materials possess several trapping levels for electrons and holes, each level being distributed around a certain activation energy (Mott and Davis 1971, Creswell *et al* 1972, van Turnhout 1975, 1980). The levels may be detected by a number of methods, such as corona charging of a one-sided metallised sample, followed by metallisation of the charged surface and subsequent charge release in a TSC experiment. Such 'short-circuit' TSC studies do not, however, readily reveal the existence of surface traps and yield only limited information on bulk traps. More knowledge is gained by the combination of this method with an 'open circuit' TSC measurement on samples not metallised on the charged surface and with electron-beam studies.

In recent experiments, von Seggern (1979) negatively corona charged 25 μm fluoroethylenepropylene (FEP) samples and depolarised them in open circuit and in short circuit. The results of a series of such open circuit measurements are shown in figure 3. The TSC

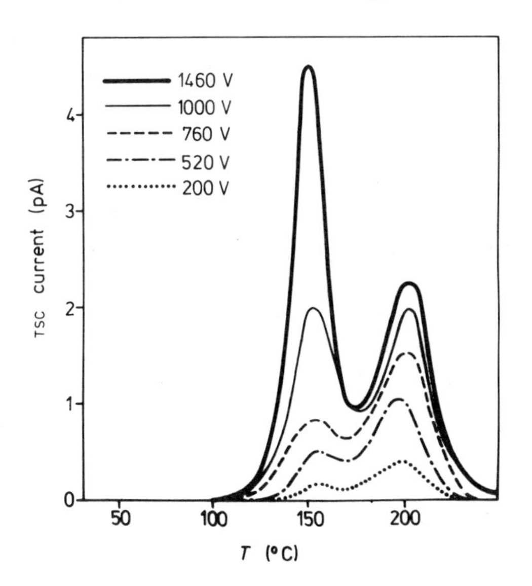

Figure 3. Open circuit TSC curves for 25 μm FEP samples charged by corona method to different potentials (von Seggern 1979).

curves, obtained from samples charged to different potentials, exhibit peaks at 155 and 200°C. On the other hand, short circuit TSC curves have peaks at 95, 170, and 200°C, but none at 155°C. Since carrier release from surface traps can only give TSC signals in open circuit experiments, the 155°C peak must be due to such traps. Similarly, the 200°C peak, which appears under open circuit and short circuit conditions, must be due to volume traps. Other reasoning shows that the 95 and 170°C peaks are caused by carrier flow in the conduction band under trap-filled limit conditions and to near-surface traps, respectively.

The geometrical location of these traps has been found from experiments with electron-beam charged samples. If samples are charged with electrons of different energies, the penetration depth and thus the depth of the initial charge location varies. Thus, open circuit TSC studies on such samples will yield the peaks due to surface and near-surface traps only for low electron-beam energies. This is seen in figure 4, where the surface peak

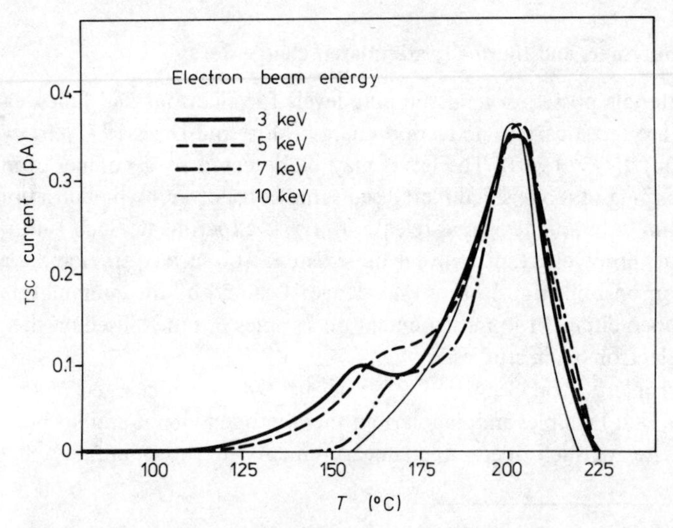

Figure 4. Open circuit TSC curves for 25 μm FEP samples charged with electron beams of different energies as indicated (von Seggern 1979).

at 155°C disappears for electron-beam energies above 5 keV and the near-surface peak at 170°C disappears for energies of 10 keV and over. Evaluation of these results with the electron-penetration data of Gross *et al* (1977) yields the trap locations shown in table 1. Also given in the table are activation energies obtained from the initial rise of the TSC curves or taken from the literature.

The TSC curves in figure 3 allow also an estimate of the densities of surface and volume traps: for surface potentials up to 760 V, both peaks rise about proportionally with the potential. If the surface potential exceeds 1000 V, the first peak rises more

Table 1. Geometrical and energetical trap distribution in 25 μm thick FEP (type A).

Peak temperature (°C)	Location relative to surface (μm)	Activation energy (eV)	Kind of traps	Density of traps
95	0–25	1.0	Energetically shallow traps under TFL conditions	unknown
155	0–0.5	1.2	Surface traps	$>10^{12}$ cm^{-2}
170	0.5–1.8	1.2	Near-surface traps	unknown
200	1.8	1.8	Bulk traps	1.4×10^{14} cm^{-3}

strongly at the expense of the second. This behaviour has been interpreted by von Seggern (1979) as follows. For samples charged to high surface potentials the first carriers activated from surface traps at temperatures around 155°C fill bulk traps within the region of their mean free path. When all such traps are filled, further carriers activated will fill bulk traps deeper in the material. Eventually, all bulk traps are occupied and carriers released from the surface now move directly to the rear electrode. Since a carrier's contribution to the current is proportional to its Schubweg (see, e.g. Sessler 1980), this explains the more than proportional increase of the first peak with surface potential. An evaluation of the curves gives the volume density of the bulk traps and a lower limit for the planar density of the surface traps. The resulting values are also given in table 1.

Other results by von Seggern (1981) show that in FEP the number of near-surface traps for positive charges is considerably greater than the number of bulk traps, although a bulk trap exists which releases holes only at about 200°C. Some of the results on storage of positive charges were obtained with the thermal pulse method by Collins (1976), which yields extremely valuable data on charge centroid locations.

Open circuit TSC data obtained on negatively charged polyethylene terephthalate (PET) is shown in figure 5. Peaks are present at 80 and 115°C regardless of the charging method.

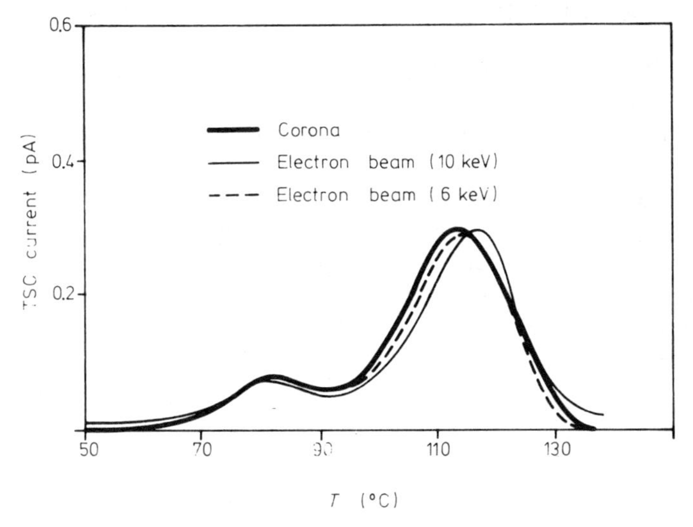

Figure 5. Open circuit TSC curves for corona charged and electron-beam charged 25 μm PET samples (von Seggern 1981).

This indicates the absence of a specific surface trap in this material. On the other hand, negatively charged polyethylene (PE) exhibits an open circuit TSC peak due to volume traps at 40°C or lower and a peak due to surface traps at above 110°C.

Based on the information gained about trapping in FEP, von Seggern (1979, 1980) has developed models for the isothermal and thermally stimulated charge decay in corona- and electron-beam-charged FEP. The models assume the existence of a surface- and a bulk-trapping level and of a trap-modulated mobility, and accurately predict the experimental results.

4. Radiation effects

Apart from its usefulness in the production of well controlled electrets, irradiation with electrons allows one to determine a number of properties of charged samples. Examples are the study of carrier mobilities, the observation of transit phenomena, and the measurement of the spatial distribution of the charge in the thickness direction of thin dielectric films. It is therefore important to understand the response of polymers to electron irradiation. Three recent studies, to be discussed in the following, have been devoted to this topic.

In an analytical investigation, Labonte (1980) has calculated the charge build-up and the current in a dielectric under electron irradiation for three boundary conditions, namely open circuit of the sample with grounded rear electrode, open circuit with grounded front electrode and short circuit. A number of different models for the electron-beam irradiation were used. These are (1) the 'box model' which assumes constant radiation-induced conductivity (RIC) in the irradiated volume and zero conductivity elsewhere (Gross *et al* 1974), (2) a model assuming a time-dependent RIC which increases according to the function $1 - \exp(-t/t_0)$, (3) a model assuming range straggling of the electrons according to the electron-deposition profiles by Matsuoka *et al* (1976), and (4) a model considering the reduction of the electron penetration depth due to space charge accumulation.

The results of these calculations for the short circuit boundary conditions are shown in figure 6. One of the interesting aspects of the figure is the fact that the current decay for model (3) is much slower than for the other models. Thus, a considerably stronger charge build-up in the dielectric, proceeding with a steadily increasing time constant, is expected. This result, which can also be derived from the work of Gross and Ferreira (1979), is of significance for the evaluation of data obtained with the electron-beam method to determine charge distributions (see below).

The slow decrease of the short circuit current under electron irradiation was recently confirmed in an experiment by Gross *et al* (1980). The results are shown in figure 7.

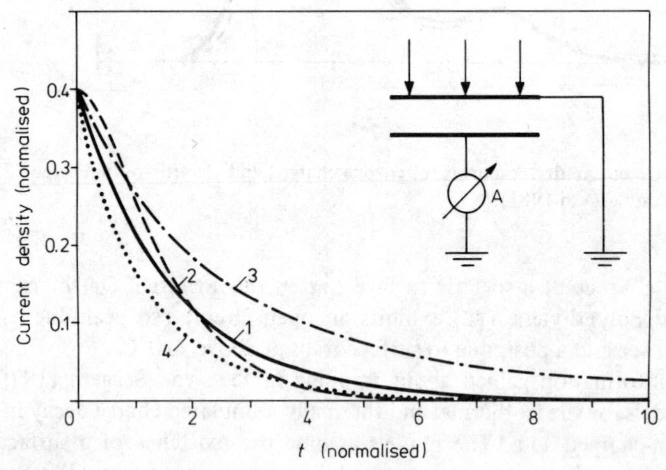

Figure 6. Normalised current due to irradiation of a dielectric as a function of normalised time. The numbers on the curves refer to the models described in the text (Labonte 1980).

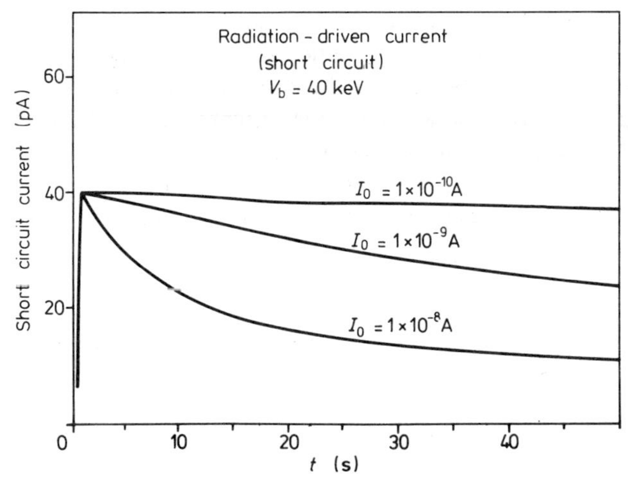

Figure 7. Measured rear-electrode currents in 25 μm FEP upon irradiation with a 40 keV electron-beam of current I_0. The currents have to be multiplied by the following factors: top curve, ×1; middle curve, ×10; bottom curve ×100. (Gross *et al* 1980).

Comparison of the data with theory indicates that for an injection current of 10^{-9} A, the rear-electrode current should drop to less than 1% of its original value within 50 s if range straggling is not considered. The absence of such a steep drop suggests the importance of range straggling of the electrons.

The main topic of Gross's work is, however, the time dependence of the RIC under continuous electron irradiation. This time dependence is assumed to be again given by $1 - \exp(-t/t_0)$. For beam densities in the range of 10^{-9} to 10^{-10} A cm^{-2}, t_0 is of the order of a few seconds. The relatively slow build-up of the RIC is clearly seen in experiments where a bias V_0 is applied across the sample, e.g. in the set-up shown in figure 8.

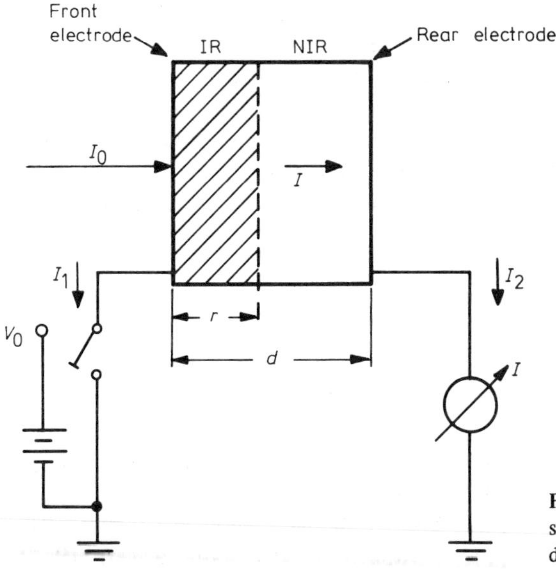

Figure 8. Schematic view of the measuring set-up for determining the response of a dielectric to irradiation (Gross *et al* 1980).

Two different experiments were performed: (1) the film was exposed to the electron beam prior to voltage application and (2) the voltage was applied prior to the beam. The measured responses are depicted in figure 9. If the beam is applied first, the RIC is fully developed at the time of voltage application and the current reaches its maximum immediately (because of the finite risetime of the recorder the maximum in the figure is reached only after about 0.5 s). If, however, the voltage is applied prior to, or simultaneously with, the beam, it takes a few seconds for the RIC to develop and the current rises slowly.

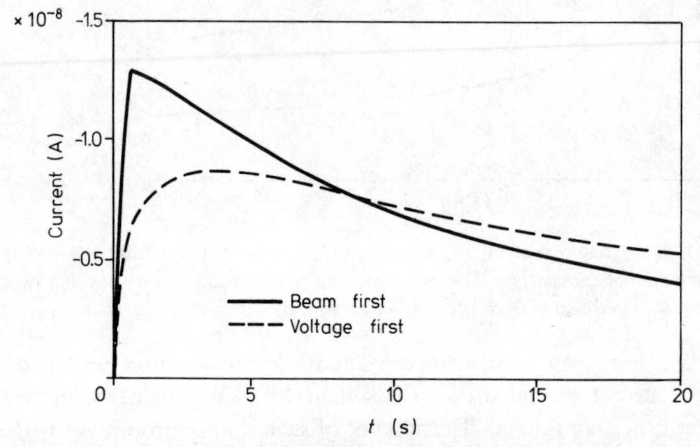

Figure 9. Induced current in a 25 μm FEP sample irradiated with a partially penetrating electron-beam. Broken curve: voltage applied before onset of irradiation. Full curve: voltage applied 10 s after onset of irradiation (Gross *et al* 1980). $V_b = 40$ keV, $I_0 = 2 \times 10^{-9}$ A, $V_0 = -300$ V.

A third experiment on radiation effects was devoted to the build-up of a polarisation in polytetrafluoroethylenc (PTFE). Von Seggern and West (1981) found that application of a voltage to a PTFE sample subjected simultaneously to penetrating radiation will yield a polarisation detectable in a TSC experiment. The results of a series of TSC experiments on 6 μm samples pretreated with 61 keV electrons and a bias of −80 V are shown in figure 10. The interesting aspect of these results is that, for long enough polarisation times, the integral over the depolarisation current very nearly reaches the value CV, where C is the sample capacitance and V the applied voltage. This suggests either the presence of a polarisation in this normally nonpolar material or the existence of space charge clouds of total charge $q_{1,2} = \pm CV$ close to the two sample electrodes, respectively, but separated from them by blocking layers.

5. Charge distribution

Several methods have been suggested to measure the charge centroid or the actual charge distribution in thin-film dielectrics. Of present interest are the thermal-pulse method of Collins (1976), the acoustic pulse method of Laurenceau *et al* (1977), and the electron-beam method of Sessler *et al* (1977). While the thermal pulse method is a convenient and accurate tool for the measurement of the mean charge depth, it can only detect certain features of charge distributions, such as a number of spatial Fourier coefficients. The distribution as such is not found uniquely (De Reggi *et al* 1978, von Seggern 1978).

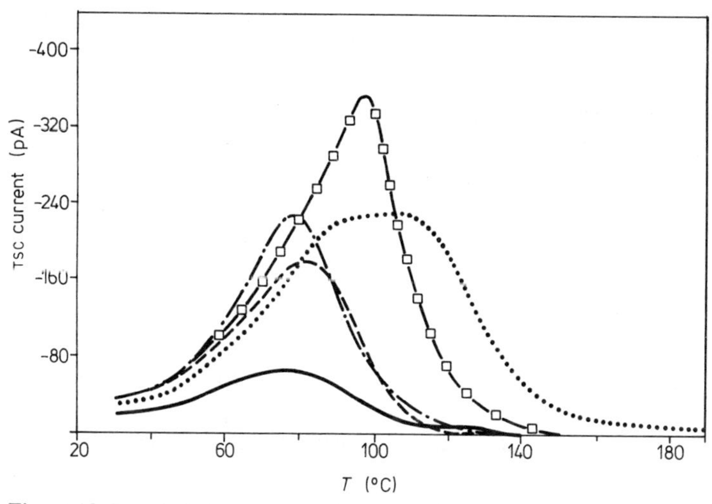

Figure 10. Depolarisation currents for various 6 μm PTFE samples. Prior to the experiment, the samples were charged by irradiation with a 61 keV electron-beam and simultaneous application of a voltage of −80 V for the following durations (the integrated charge from the depolarisation current is also given): ——, 10 s, 0.2 *CV*; – – – –, 50 s, 0.5 *CV*; – · – · –, 100 s, 0.56 *CV*; □–□, 500 s, 0.9 *CV*; . . . . . . , 1000 s, 0.96 *CV*: (von Seggern and West 1981).

The acoustic pulse method determines the distribution uniquely but is presently limited by a spatial resolution of 10 to 100 μm (Laurenceau *et al* 1977, Migliori and Thompson 1980). The electron-beam method, although destructive, is capable of generating unique and high-resolution data in certain materials (see Sessler 1980). Recently this method has been applied to determine the charge distribution in PET with a resolution of about 1 μm (Tong 1980, Sessler *et al* 1981).

The electron-beam method is based on an evaluation of induction charges released from the rear electrode of a charged sample during irradiation through the front electrode with a monoenergetic beam of electrons. The method is schematically described in figure 11. If the energy of the electron-beam is increased the zone of RIC, whose front can be considered a virtual electrode, is extended and more charges are released from the rear

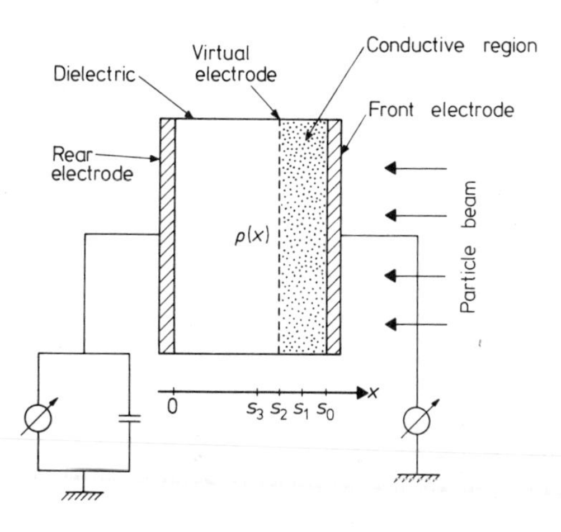

Figure 11. Schematic diagram of the set-up for measurement of the charge distribution in a thin film dielectric (Sessler *et al* 1977).

electrode. For a continuous motion of the virtual electrode, an evaluation of the rear-electrode charge q yields the charge distribution $\rho(x)$ by means of

$$\rho(x) = -\mathrm{d}^2\,(qx)/\mathrm{d}x^2. \qquad (1)$$

For step motion, this equation is replaced by a finite-difference relation. A calibration run on an uncharged sample is necessary to eliminate the effect of charges deposited by the probing beam. It can be shown that the calibration is only needed because of the range straggling effect discussed above and would not be required if the simple box model applied.

A number of basic experiments were recently performed to determine the feasibility and the validity of the method as applied to PET (Sessler *et al* 1981). In some of these experiments, the fraction of the originally trapped charge that can be removed from the sample with this method was measured. The fraction depends, of course, on the time and intensity of irradiation but is fairly independent of the energy of the irradiating beam as long as the virtual electrode penetrates beyond all trapped charges.

A typical result of these experiments is shown in figure 12. Here, the 25 μm PET sample was first charged with a 25 keV electron beam. The charge was then 'recalled' with a 30 keV beam of density 10^{-9} A cm^{-2} which generates a virtual electrode beyond all the charges deposited by the 25 keV beam. The measured dependence on irradiation time shown in the figure is such that ~90% of the original charge is removed after a period of about 300 s.

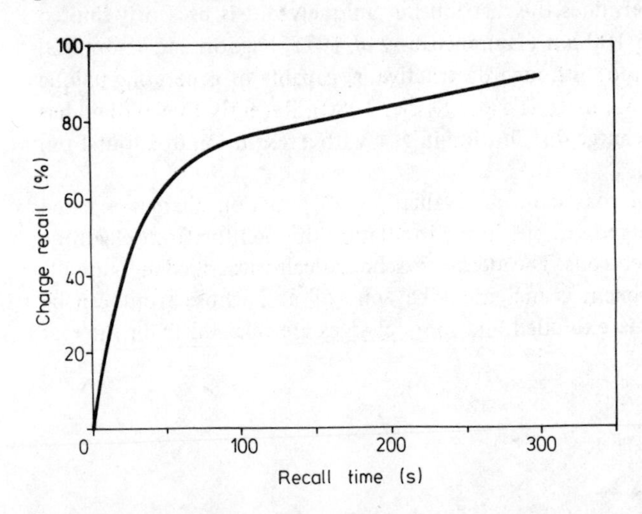

Figure 12. Percentage of charge 'recalled' from a 25 μm PET sample previously charged with a 25 keV electron-beam. The probing beam had an energy of 30 keV and a current density of 10^{-9} A cm^{-2} (Sessler *et al* 1981).

Experiments were also performed to determine the location of the virtual electrode and its independence of the electric field existing in the sample. Figure 13 shows results on the location of the virtual electrode at fields up to about 100 kV cm^{-1}, where a field dependence does not exist. The experiments were performed by applying a positive voltage V to the front electrode of the sample. Upon irradiation of this electrode, a virtual electrode is formed at a depth ds in the dielectric and a charge dQ = (ds/s)CV, where C is the sample capacitance, is transferred in the circuit. A measurement of dQ thus yields ds.

Also shown in figure 13 are results for the centroid locations of charges deposited by electron beams under open and short circuit conditions. These results were obtained with

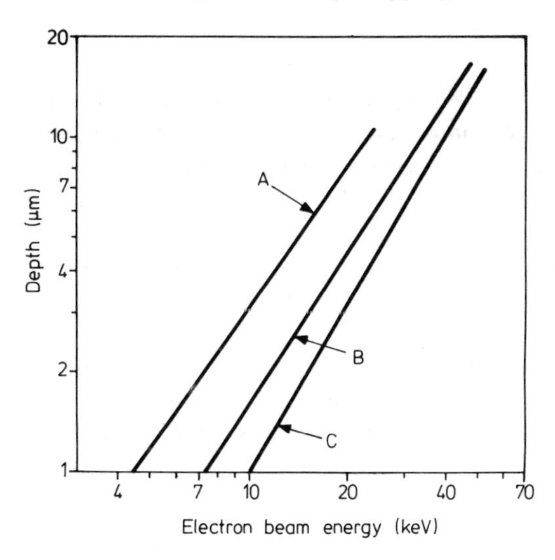

Figure 13. Location of the virtual electrode (preliminary results) and the mean depth of charge deposited by an electron-beam under open and short circuit conditions in 25 µm PET. A, virtual electrode; B, open circuit charge centroid; C, short circuit charge centroid; (Sessler *et al* 1981).

the split Faraday cup method by Gross *et al* (1977) on samples positively biased with respect to the sample holder to avoid secondary electron emission. Comparison of the data in the figure reveals that the virtual electrode is about twice as deep as the centroid of the charge deposited under open circuit conditions. This is reasonable in view of the charge-deposition data by Beers *et al* (1978) and Berkley (1979) for electron-beam irradiated polymers.

Charge-distribution measurements with this method were carried out on 25 µm PET samples charged by electron bombardment prior to the experiment. The results obtained for a sample charged with 20 keV electrons are shown in figure 14. Since according to equation (1) a second derivative has to be used for the evaluation, the errors in $\rho(x)$ are relatively large; they are estimated to amount to about 2 relative units. Thus, only the main

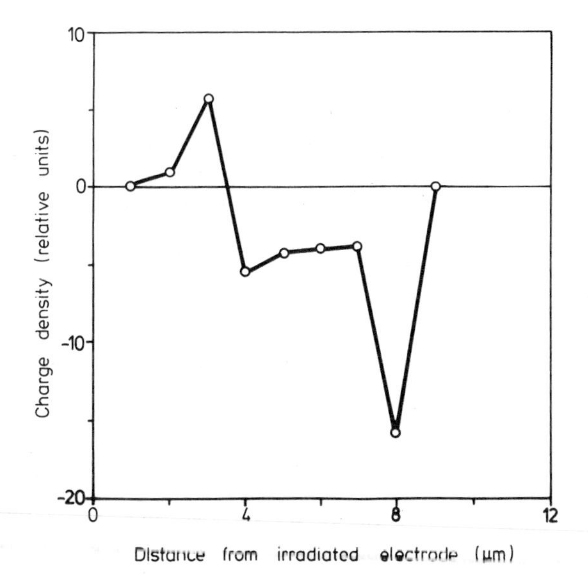

Figure 14. Charge distribution in a 25 µm PET sample, charged with a 20 keV electron-beam prior to the distribution measurement (Sessler *et al* 1981).

features of the distribution should be considered. A step size of 1 μm for the virtual electrode was achieved by proper selection of electron-beam energies.

Surprisingly, a small positive charge is found close to the front electrode (left in the figure) while the negative charge is embedded more deeply in the material. No charge was detected at depths in excess of 9 μm. The positive charge probably results from the injection of carriers left on the sample electrode due to secondary emission (this sample was not biased during charging). The amount of positive charge (about 15% of the negative charge) is certainly explicable from the secondary emission yields of the aluminium electrode. Due to the positive charge, the charge centroid is deeper than expected from figure 13.

Since this method is the first one capable of sampling charge distributions with a resolution of about 1 μm its application to other electrets and to photoconductors, metal oxides, etc, might furnish valuable information about charge storage and charge decay phenomena in these materials.

6. Conclusions

A number of recent studies have furthered the understanding of charge-trapping and charge-decay phenomena in polymers. In FEP, which is important because of its excellent charge-storage capabilities, new results about the trap structure have aided in the development of successful models for the isothermal and thermally stimulated charge decay of surface charges and bulk charges. Also, a more satisfactory picture of radiation phenomena has been achieved which is vital for the use of diagnostic tools depending on irradiation techniques. Further progress towards a more comprehensive description of charge trapping depends on the capability of determining the distribution of charges in the direction of thickness of thin polymer films. A first step in this direction is a new electron-beam probing technique which has already furnished promising results in PET. Further use of this method might lead to a more complete understanding of charge-trapping phenomena in polymers.

Acknowledgments

The author is grateful to his coworkers at the Technische Hochschule Darmstadt and at Bell Laboratories, Drs D A Berkley, R Gerhard, B Gross, K Labonte, H von Seggern, and J E West, for permission to use some of their results as well as joint data prior to publication. Funding of part of this research by the Deutsche Forschungsgemeinschaft is also acknowledged.

References

Beers B L, Hwang H, Lin D L and Pine V W 1978 *Spacecraft Charging Technology* (NASA Conference Publication 2071) pp209–38

Berkley D A 1979 *J. Appl. Phys.* **50** 3447–53

Collins R E 1976 *J. Appl. Phys.* **47** 4804–8

Creswell R A, Perlman M M and Kabayama M A 1972 *Dielectric Properties of Polymers* ed. F E Karasz (New York: Plenum) pp295–312

De Reggi A S, Guttmann C M, Mopsik F I, Davis G T and Broadhurst M G 1978 *Phys. Rev. Lett.* **40** 413–16

Gerhard R 1981 to be published
Gross B and Leal Ferreira G F 1979 *J. Appl. Phys.* **50** 1506–11
Gross B, Sessler G M and West J E 1974 *J. Appl. Phys.* **45** 2481–51
— 1977 *J. Appl. Phys.* **48** 4303–6
Gross B, West J E, von Seggern H and Berkley D A 1980 *J. Appl. Phys.* **51** 4875–81
Ieda M, Sawa G and Shinohara U 1967 *Jap. J. Appl. Phys.* **6** 793–4
Labonte K 1980 *Ann. Rep., Conf. Electrical Insulation and Dielectric Phenomena* (Washington, DC: NAS) pp321–7
Laurenceau P, Dreyfus G and Lewiner J 1977 *Phys. Rev. Lett.* **38** 46–9
Matsuoka S, Sunaga H, Tanaka R, Hagiwara M and Araki K 1976 *IEEE Trans. Nucl. Sci.* **NS-23** 1447–52
Migliori A and Thompson J D 1980 *J. Appl. Phys.* **51** 479–85
Moreno R A and Gross B 1976 *J. Appl. Phys.* 47 3397–402
Mott N F and Davis E A 1971 *Electronic Processes in Non-Crystalline Materials* (Oxford: Clarendon Press)
von Seggern H 1978 *Appl. Phys. Lett.* **33** 134–7
— 1979 *J. Appl. Phys.* **50** 2817–21, 7039–43
— 1980 *Ann. Rep., Conf. Electrical Insulation and Dielectric Phenomena* (Washington, DC: NAS), pp345–52
— 1981 to be published
von Seggern H and West J E 1981 to be published
Sessler G M (ed.) 1980 *Electrets* (Berlin: Springer)
Sessler G M, von Seggern H and West J E 1981 to be published
Sessler G M and West J E 1974 *Electrophotography, Second Int. Conf.* ed. D R White (Washington, D C: Soc. Phot. Scient. and Engineers) pp162–6
Sessler G M, West J E, Berkley D A and Morgenstern G 1977 *Phys. Rev. Lett.* **38** 368–71
Tong D 1980 *Proc. IEEE Int. Symp. Electrical Insulation* (New York: IEEE) pp179–83
van Turnhout J 1975 *Thermally Stimulated Discharge of Polymer Electrets* (Amsterdam: Elsevier)
— 1980 *Electrets* ed. G M Sessler (Heidelberg: Springer) pp81–215

Appendix: list of contributed papers

The assessment of dielectric information

R M Hill and L A Dissado
Chelsea College, University of London, Pulton Place, London SW6, UK

Low temperatures and low-frequency dielectric relaxation in some organic glasses

L Jorat, G Novel, A Bondeau and J Huck
Laboratoire de Physique de Materiaux, 42023 St Etienne Cedex, France

Automatic computation of the distribution function of relaxation frequencies from dielectric spectra

Y Balcou
Départment de Physique Cristalline et Chimie Structural, Université de Rennes, Campus de Beaulieu, 35031 Rennes Cedex, France

Dielectric properties of ionic conductors: three compounds, a similar behaviour

P Abelard, J F Baumard and B Cales
Centre de Recherche sur la Physique des Hautes Températures, CNRS, 45045 Orléans Cedex, France

The effect of pressure on dielectric relaxation in the alkaline earth fluorides

J Fontanella, M C Wintersgill, A V Chadwick and C Andeen
Physics Department, Department of the Navy, United States Naval Academy, Annapolis, Maryland 21402, USA

The study of dielectric properties of biologically important compounds – glucose, galactose and lactose

M Mahajan, V A Garg and K N Saxena
Department of Physics, University of Indore, Indore (MP) 452001, India

Noise effects in dielectrics

L A Dissado and R M Hill
Department of Physics, Chelsea College, London University, Pulton Place, London SW6, UK

Crystalline dielectric relaxation processes in poly(vinylidene fluoride) in forms I and II

A R Blythe, G R Davies and A J Killey
ICI Plastics Division, Welwyn Garden City, Herts, UK

Relaxations in the quantum limit

R Isnard and J le G Gilchrist
Centre de Recherche sur les très Basses Températures, CNRS BP-66X,
38042 Grenoble Cedex, France

Low-temperature dielectric behaviour of γ-alumina

G P Singh, M von Schickfus, S Hunklinger and K Dransfield
Max Planck Institut für Festkörperforschung, Stuttgart,
Germany

Behaviour of liquid helium facing the sub-microscopic porosity of some solid dielectrics

J M Goldschvartz
2287 Rijwijk (ZH), The Netherlands

Theoretical and experimental studies of high-frequency dielectric relaxation

J McConnell, J R Birch and G W Chantry
Dublin Institute for Advanced Studies, Burlington Road,
Dublin 4, Eire

A critical point analysis of multiphonon structure in far-infrared dielectric response of some II–VI and III–V compound semiconductors

T J Parker, A Memon, D J Bradshaw and J R Birch
Department of Physics, Westfield College, University of London, Kidderpore
Avenue, London NW3, UK

FIR matrix spectroscopy at very low temperatures

F Hufnagel
Institut für Physik, University of Mainz, Mainz,
Germany

A novel photo-dielectric cross effect

E E Havinga
Philips Research Laboratories, Eindhoven, The Netherlands

Dynamic methods for determination of the charge density on unipolar electrets

P I Kuindersma and R M van der Heij
Philips Research Laboratories, Eindhoven, The Netherlands

Surface charge distribution of corona charged electrets

R Gerhard
Institut für Elektroakustik, Technische Hochschule, Merckstrasse 25,
D-6100 Darmstadt, Germany

Insulating layer and interfacial recombination: their joint effect in MIS solar cells

P T Landsberg and C M H Klimpke
Faculty of Mathematical Studies, University of Southampton, Southampton, SO9 5NH, UK

Study of changes of phases in poly(vinylidene fluoride)

M Latour and G Geneves
Laboratoire de Physique Moleculaire et Cristalline, Université des Sciences et Technique du Languedoc, Montpellier, France

Spontaneous electric polarisation: a phenomenon always present in the dielectric solids just prepared

D Bertolini, M Cassettari and G Salvetti
Consiglio Nazionale delle Richerche, Laboratorio di Fisica Atomica e Moleculare, 56100 Pisa, Italy

Molecular motions in solid polysaccharides by TSC

K Nishinari, D Chatain and C Lacabanne
Laboratoire de Physique des Solides, Université Paul-Sabatier, Toulouse, France

Electric field dependence of dielectric tangent loss at microwave frequencies in displacive ferroelectrics with impurities

B S Semwal
Garhwal University, Srinagar, Garhwal UP, India

Dielectric properties of new non-stoichiometric ferroelectric phases appearing in the ternary diagram Li_2O–Ta_2O_5–TiO_2

B Elouadi, N Zriouil, R von der Mühll, J Ravez and P Hagenmuller
Department of Chemistry, Faculty of Sciences, University of Rabat, Rabat, Morocco

Mechanical experiments in the study of dielectric phenomena

E R Moganischi, G B Parravicini and A Savini
Istituto di Fisica Generale 'A. Volta', Universita di Pavia, I-27100 Pavia, Italy

Some new results concerning the piezoactivity of PVDF electrets

P T A Klaase and J van Turnhout
Division of Technology for Society, Organisation for Industrial Research, TNO, PO Box 217, 2600 AE Delft, The Netherlands

Polarisation effects in gamma and electron beam irradiated polytetrafluoroethylene

H von Seggern and J E West
Bell Laboratories, 600 Mountain Avenue, Murray Hill, New Jersey 07974, USA

Dielectric spectroscopy of curing polyurethane

G Arlt and G Babiel
Institut für Werkstoffe der Elektrotechnik der Rhein, Technische Hochschule Aachen, Templergraben 55, 5100 Aachen, Germany

The dielectric properties of polycrystalline ice – dependence on the purity of water used

P Pissis and L Apekis
Physics Laboratory A, National Technical University, Zografou Campus, Athens 624, Greece

Electrostriction in perovskite crystals

L E Cross, R E Newnham, K Uchina and S Nomura
Materials Research Laboratory, Pennsylvania State University, University Park, PA 16802, USA

The dielectric response of solids at very low temperatures

R M Hill and A K Jonscher
Department of Physics, Chelsea College, University of London, Pulton Place, London SW6, UK

Deformation activated charging and discharging of polymers

J van Turnhout and P T A Klaase
Division of Technology for Society, Organisation for Industrial Research, TNO, PO Box 217, 2600 AE Delft, The Netherlands

The dielectric response of p–n junctions

A K Jonscher, V Charoensiriwatana, J Favaron and C K Loh
Department of Physics, Chelsea College, University of London, Pulton Place, London SW6, UK

Unipolar injector regimes in the current–voltage characteristics of insulator thin films

P Dellanoy
Groupe de Physique des Solides de l'ENS, Université de Paris VII, 2 Place Jussieu, 75221 Paris Cedex 05, France

X-ray induced currents in poly(vinylidene fluoride)

M Ieda, T Mizutani, V Suzuoki and N Sugiua
Department of Electrical Engineering, Nagoya University, Furo, Chikusa, Nagoya 464, Japan

Isothermal and thermally stimulated current studies of positive corona-charged Teflon FEP

H von Seggern and J E West
Bell Laboratories, 600 Mountain Avenue, Murray Hill, New Jersey 07974, USA

Bulk and contact contributions in nylon 66 dielectric response

N Rosenberg and M Maitrot
Université Lyon I, 43 Bd 11 Novembre 1918, 69622 Villeurbanne, France

Effects of poling field and time on pyroelectric coefficient and polarisation uniformity in polyvinylfluoride

S B Lang
Department of Chemical Engineering, Ben Gurion University of the Negev, Beersheva, Israel

RANDALL LIBRARY-UNCW
3 0490 0090962 %